TR

The Rise and Fall of the British Trawl Fishery

From reviews of the hardback edition of this book:

"... the most valuable contribution to the literature so far. It should be read by everyone who claims a say in fisheries policy matters and also as a case study of the rise and fall of a major industry under the special influences of a common-property resource, governmental management and international law."
The Northern Mariner

"Robinson's treatment is sufficiently academic to satisfy the specialist but also an absorbing read for the layman."
The Maritime Yearbook

Robb Robinson was born into a Hull seafaring family which has been involved in fishing and maritime commerce for generations. He has written extensively on the fishing industry for a variety of national and international publications and is a member of staff at Hull College.

The cover illustration is from a painting called "Fleeting" attributed to Kenneth Hote c.1890 and is reproduced by kind permission of the National Fishing Heritage Centre, Grimsby.

EXETER MARITIME STUDIES

General Editors: Michael Duffy and David J. Starkey

British Privateering Enterprise in the Eighteenth Century
by David J. Starkey (1990)

Parameters of British Naval Power 1650–1850
edited by Michael Duffy (1992)

The Rise of the Devon Seaside Resorts, 1750–1900
by John Travis (1993)

Man and the Maritime Environment
edited by Stephen Fisher (1994)

Manila Ransomed: The British Assault on Manila in the Seven Years War
by Nicholas Tracy (1995)

Pirates and Privateers: New Perspectives on the War on Trade in the Eighteenth and Nineteenth Centuries
edited by David J. Starkey, E.S. van Eyck van Heslinga and J.A. de Moor (1997)

Cockburn and the British Navy in Transition: Admiral Sir George Cockburn, 1772–1853
by Roger Morriss (1997)

Recreation and the Sea
edited by Stephen Fisher (1997)

Exploiting the Sea: Aspects of Britain's Maritime Economy since 1870
edited by David J. Starkey and Alan G. Jamieson (1998)

TRAWLING

The Rise and Fall of the British Trawl Fishery

Robb Robinson

UNIVERSITY
of
EXETER
PRESS

First published in 1996 by
University of Exeter Press
Reed Hall, Streatham Drive
Exeter, Devon EX4 4QR
UK

British Library Cataloguing in Publication Data
A catalogue record for this book is
available from the British Library

Hardback edition 1996
ISBN 0 85989 480 0

Paperback edition 1998
ISBN 0 85989 628 5

Typeset in 12/13pt Linotype Garamond 3
by Kestrel Data, Exeter

Printed and bound in Great Britain by
Short Run Press Ltd, Exeter

Contents

List of Tables

List of Maps and Charts

List of Figures

List of Plates

List of Abbreviations

BPP	British Parliamentary Papers
HCRO	Humberside County Record Office
HRO	Hull Record Office
NEDSFC	North Eastern District Sea Fisheries Commitee
NYCRO	North Yorkshire County Records Office
PRO	Public Records Office
RC	Royal Commission
RHE	Register House, Edinburgh
SC	Select Committee

Introduction

In late January 1955 the north-west coast of Iceland was gripped by the worst blizzard in memory. Deep-sea trawling fleets from several nations had run for the relative shelter of Isafjord and neighbouring anchorages as snow, ice and seventy-five mile-per-hour gales whipped the Arctic seas into a grim grey and white hell. Yet one vessel could not make shelter. The Hull trawler, *Kingston Garnet*'s fishing gear had fouled her propeller leaving her at the mercy of this ice-cold inferno. Her distress calls were soon picked up by fellow Hull trawler, *Lorella*.

Such was the fellowship of the trawling tradition in the face of elemental danger that the *Lorella*'s skipper, Steve Blackshaw, was soon heading out of his safe anchorage and into the worst possible of weathers beyond Iceland's North Cape in a brave attempt to aid his stricken colleagues. Before long the *Lorella* was also in deep trouble, overwhelmed by the phenomenon sometimes known as black frost. Haze and spray, everything froze as it hit the ship and ice swiftly built up on the trawler's rigging and superstructure making her dangerously top heavy. The black frost froze everything in its path; a half-inch diameter wire rope might gain a six-inch coating of ice and on some occasions trawlermen had to chop fish from the nets as they came on deck because they had frozen together. Usually the ice could be chipped off before it became too dangerous but this time the build-up on the *Lorella* was so unexpectedly fast that the crew's frantic efforts to hack it off were to no avail. Radio operator, George Hobson gave out a desperate SOS 'Have been thrown on side need quick help'.

A third Hull ship, *Roderigo*, skippered by Steve Blackshaw's friend, George Coverdale, picked up this desperate call and raced out into the wild seas. Within a short while she too was stricken by ice. Other skippers, listening on their radios, heard the two vessels speaking to each other throughout their final ordeal. The *Lorella* was the first to capsize and the quiet voice of Steve Blackshaw closed with 'I am going over now. I cannot abandon ship. Mayday!' A little later George Coverdale's vessel also slipped over whilst putting out the same Mayday call.

The *Kingston Garnet* managed to free her propeller and limp to safety. Aircraft joined in the search for survivors but the other two crews stood little chance in the ice-cold water. Further round the coast, some twenty miles west of the North Cape the weather had claimed another trawler: the Icelandic vessel *Egill Raudi* was reported on her side on the south shore of Cape Ritur. The *Lorella*'s sister ship, *Lancella*, refused to be beaten by the conditions and continued to search, despite the atrocious weather, for eighteen hours after the final distress calls had been picked up. Eventually she too had to give up hope that any of the forty men would be found alive.

The *Lorella* and *Roderigo* are just two out of thousands of British trawlers that have been lost over the years. Hull has lost 900 alone. In 1969 it was calculated that a trawler deckhand worked with a three times greater risk of death than a miner at the coalface. Skippers and mates, working longer hours under greater stress, were even worse off and the risk to their lives was considered to be twenty times greater than any other industrial risk in the world. Trawlermen worked—indeed still work—in conditions which on land were consigned to the history books by nineteenth-century factory reformers. Deep-sea trawlermen were—and are—used to working in almost all weathers and accepted as normal an eighteen-hour day on the fishing grounds. Death and injury seemed almost as commonplace in the 1950s as they had been a century before.

In the middle of the twentieth century the British deep-sea trawling trade was at its peak. Trawlers not only fished off Iceland and the Norwegian coast but also visited Bear Island, the Barents Sea and Greenland amongst other places. Grimsby and Hull were the largest fishing ports in the world. But within thirty-five years of the loss of the brave crews of the *Lorella* and *Roderigo* the great distant-water sector of the British trawling fleet had all but disappeared. British—or more correctly European—rather than Arctic waters are now the main hunting ground of a restructured British fishing fleet. The distant-water trawlers did not succumb to those formidable Arctic elements which claimed so many individual vessels and their crews, but rather to the combined forces of international politics and conservation. A unique way of life has gone with this fishery and the aim of the following chapters is to chronicle its rise and fall.

The rise of the British trawl fishery is essentially a story of the nineteenth and twentieth centuries, but fishing in one form or another has provided man with both food and a livelihood since prehistoric times. Throughout the ages fishermen have had to contend with the violence of storm, wind, water and sometimes other men in their attempts to wrest a livelihood from the sea. Such drama has sometimes led the casual observer to romanticize their calling but nothing could be further from the truth. Fishing has always been one of the most dangerous and uncertain occupations on earth, requiring excessively long and unsociable hours of hard, cold toil. Generations of deep-sea fishermen became accustomed to spending weeks on grey watery wastes, far away from family and friends, in conditions the landsman can scarcely imagine. A fisherman's calling is unique and escapes easy definition: in reality it cannot be simply described as either orthodox industry, agriculture or transport but contains elements of all three. Fishermen have always been hunters and are today amongst the last groups who hunt for a living in the western world. Small wonder their story provides a unique chapter in Britain's economic and social history.

A deep-sea trawlerman can simply be defined as one who goes out into deep water to trawl. This book is primarily concerned with the leading edge of the activity, the distant-water trawling trade. By the mid-twentieth century the deep-sea fishing industry was often considered to consist of three sectors: the near-water, middle-water and distant-water fleets. After the Second World War the near- and middle-water sectors were made up of decked vessels up to 140 feet in length that fished principally on grounds around the Faroes, in the North Sea, the English Channel, or off the west coast of the British Isles. The distant-water fleet mainly consisted of vessels exceeding 140 feet in length which made trips to distant-water grounds off Iceland, Bear Island and the like.

In reality the fishing industry has always been far more complex. First of all there is the difference between the pelagic and demersal fisheries. For many years fishermen often specialized in the pursuit of one or the other and the white fish and herring trades, for example, were for a long time overseen by separate government bodies. Different fishermen used different gear and white fish were taken by a variety of methods including trawling, lining and seine netting. In the early nineteenth century, line fishing predominated until superseded by trawling. By the later 1950s seining was growing in popularity. Moreover, the definition of what could be called distant-water fishing changed over the years. In the mid-nineteenth century, when the smack fishermen from the south and west were opening up trawling grounds around the Dogger Bank, many contemporaries regarded them as fishing in distant waters. But as trawling spread further to the north so the concept of what was distant water moved with them.

Space limitations require specialization and this book will throughout be primarily concerned with the evolution of the leading sector of the distant-water trawling trade: from the age of the Brixham and Barking sailing smacks to the era of the stern freezer trawler.

1

The Pre-Trawling Era

Even in the Middle Ages fishing was an important part of the economy of those European cities and states which bordered the North Sea, Baltic and Atlantic. From the early fifteenth century the Dutch were to lead the way in developing a complex trade based on the North Sea herring fisheries. Although somewhat in their shadow, the British pursued a great variety of fish, not only in home waters but soon as far afield as the Icelandic and Newfoundland coasts. By the late eighteenth century, when trawling had yet to make a substantial mark, they had developed an extremely diverse and sophisticated trade.

One of the most important of the national fisheries was that devoted to herring. They shoaled at different seasons of the year off the various coasts of the British Isles but almost all were taken by the age-old method of drifting fleets of nets from boats. There were marked regional differences of scale, intensity of effort and organization. The premier English herring fishery had long been based on Yarmouth.[1] The autumn season there usually stretched from the end of September to early December and supported a large fleet of local and visiting craft. Much of the catch, if not eaten locally or immediately shipped off to accessible towns and villages, was turned into either red or white herring.[2]

Reds were smoked over a wood-shaving fire in a smokehouse for up to twenty-one days. White or salt-cured herrings were split, salted and then packed in barrels between layers of salt. After either process the cured fish remained edible indefinitely, an essential requirement in those days of slow and uncertain travel.

Red and white herring were shipped off to the West Indies to provide food for the slaves on the sugar plantations, but after the abolition of slavery in the British Empire in 1834 those freed had perhaps more choice in their diet and the trade speedily diminished. British curing standards in the eighteenth and early nineteenth centuries were generally inferior to those of the Dutch[3] who supplied the most lucrative markets, though British white herrings were sold to Ireland whilst reds were often exported to Genoa, Venice, and the German states.

Liverpool was also an important centre for the herring trade. Fast cutters and sloops sailed from the Mersey for grounds off the Isle of Man, Cornwall, Wales or the west coast of Scotland: wherever the herring were shoaling. These vessels did not fish but bought the fresh catches of local boats each morning which were then salted and stored in barrels. A full cargo could be loaded in between fifteen and twenty-one days and on arrival back at Liverpool the fish were either smoked as red or salted as white according to demand.[4] The port was well-placed between the various fisheries and had established trading connections with inland and overseas markets as well as access to supplies of Cheshire salt.

Not all west-coast herring fishing was linked to the Liverpool trade. In Scotland the British Fisheries Society, founded in 1786, developed a series of fishing villages and harbours at Ullapool, Tobermory and Skye.[5] Vessels from the Clyde burghs also exploited the west-coast shoals and formed the basis of an important trade. Around the Cornish coast fishermen from St Ives and district sent their herrings to market by coaster in 'casks and puncheons of every description'.[6]

Other British herring seasons worked at this time included the Eastern Channel fishery, which drew craft from ports such as Portsmouth, Dover and Hastings. Then there was the Northumberland season which lasted from June to August—though this was of only local significance. Herrings shoaled off the Yorkshire coast mainly in August and September but the fishery there retained only a vestige of its medieval significance when Continental herring merchants had flocked to Scarborough and Whitby.[7]

On the east coast of Scotland herring fishing was mainly concentrated on the Firth of Forth in the 1790s but during the nineteenth century the activity was taken up by villages up and down the seaboard in order to satisfy the growing demand for salt-cured herring which, in later years, were exported to eastern Europe in huge quantities. The development of this trade was fostered by the Government's Fishery Board which started life in 1809 as the Commission for the Herring Fishery and which was responsible for raising the quality of the cured product. The English east-coast herring fishery was also to expand during the nineteenth century, especially after the railways opened up new inland markets for fresh and lightly cured herring, most notably the Newcastle kipper and Yarmouth bloater. Still later, the English fisheries also helped to satisfy the massive eastern European demand for salt-cured herring, and Scottish curers with their gangs of herring lasses, coopers and rullymen came south each year in pursuit of the drifter fleets which voyaged from one herring season to another.

In the late eighteenth century Britain also possessed an important white fishing trade. The main means of capture then used varied from hand lines to the so-called great or long lines that might stretch for up to two hundred fathoms across the sea bed and have numerous hooks attached by means of short cords known as snoods. Whatever the gear, the principle was the same in that the fish were lured to their fate by the baited hook. However, other, less important, methods were sometimes deployed: along the Yorkshire coast turbots were taken by means of stationary nets and, as we shall see below, a limited amount of trawling was practised in certain districts.

London was then the principal market for white fish, being supplied by Barking, Gravesend and Harwich men in particular. Their craft often had large flooded holds, knows as wells, which enabled them to keep much of their catch, caught out on the North Sea grounds, alive until they landed at Billingsgate.[8] The Harwich men worked on the following annual pattern which differed slightly to that of the other ports. From about June or July to early November they went hand-lining for haddock and small cod about fifteen leagues from the Norfolk and Lincolnshire coasts, after which they sought cod off the Dogger Bank with great or long lines until April. For the next month or so most of the fleet would return to the Norfolk and Lincolnshire coasts, whilst the remainder would fish for lobsters. At the end of May all would refit for another yearly round.[9]

Some craft from these ports and also Yarmouth still ventured to

Iceland but such voyages were less common than they had been in earlier centuries when many other ports, including Scarborough and Whitby, had also sent fleets there. It seems that losses of craft during the seventeenth-century Dutch wars and competition from the Newfoundland grounds had stifled British exploitation of the Icelandic fisheries, but in the nineteenth century interest was to revive. Newfoundland had been discovered after John and Sebastian Cabot sailed from Bristol in 1497 in a quest for the North West Passage to India and the rich fisheries found off its foggy coasts soon attracted West Country fishermen along with the French and Portuguese. They sailed there each spring and spent the summer hand-lining for cod, which were then split, salted and dried onshore, before being shipped home as cargo in the autumn. By the late eighteenth century the Newfoundland Grand Banks had also become an important source of dried cod for the English and Continental markets.[10]

Other British regions had a considerable financial interest in white fishing. On the Yorkshire coast, for example, the communities of Scarborough, Staithes, Flamborough, Runswick and Robin Hood's Bay owned fleets of three-masted decked luggers, measuring over fifty feet from stem to stern. Every spring these would be fitted out to follow the great line fishery for cod and ling off the Dogger Bank. The lines were not laid directly by the luggers but from cobles which they carried to sea on their decks. Come September they would fit out with drift nets and set sail for the Yarmouth autumn herring fishery, as they had done since at least the middle of the seventeenth century. When the boats returned home at the end of November they were laid up for the winter in the harbours of Whitby, Scarborough or Bridlington Quay whilst their crews returned to the inshore coble fishery.[11]

A portion of the cod and ling landed on the Yorkshire coast was dried and exported through London to Spain, the Mediterranean and the West Indies. Dry curing was a complex procedure. The fish were first split, salted and spread out on the rocks and hills by the shore until seemingly dry. Then they were collected into one large pile and left to stand for ten to twelve days: a process known as sweating. Afterwards, the stack was opened up and the fish once more exposed to the sun and air until thoroughly cured.[12] To the modern British consumer such products might seem most unappetising, but their eighteenth-century counterparts lived in an age before technological developments in the fields of transport and refrigeration revolutionized the world of food. Their tastes were quite different and good quality dry-cured fish was a favourite dish. Even today, it finds great favour in parts of southern Europe and the rest of the world. Similar dry-curing activities were to be found all round the British coasts and the Shetland Isles were the largest eighteenth-century producers, processing large quantities, especially for export to Ireland.

There were many other fisheries—too numerous to mention—and these were worked with differing degrees of intensity. Some were unique and had adapted to specific local conditions. In the Humber estuary, for example, small vessels from the townships of Patrington Haven and Paull netted shrimps or prawns which were boiled on board before being sold in local towns and villages. Shell fish, especially cockles and mussels, had been taken from the Wash since time immemorial and their beds were under the control of the corporations of Boston and Lynn.[13] Oysters—then an extremely cheap and popular dish amongst the poor—were dredged up from the sea bed at Whitstable and other places. Almost every district had crab and lobster seasons and the mackerel fisheries in the Channel provided much employment for ports such as Hastings.[14] Yet another, perhaps more crucial source of sustenance, was the pilchard fishery. This was carried out along the north coast of Cornwall by St Ives boats and on the southern coast in Mounts Bay and eastwards towards St Mawes and Mevagissey. In the mid-1780s hundreds of small boats went out seining for pilchards whilst countless other shoals were caught by inshore nets hauled on to the beach. On shore, some four or five thousand people found employment processing and packing the fish or else pressing them to obtain oil. A great deal were exported, especially to Madeira, the West Indies, Ostend and the Mediterranean,[15] but they were also an essential part of the local diet and, for many poor families, pilchards and potatoes provided the principal means of winter nourishment.

Despite the notorious difficulties of transporting anything overland in the pre-railway era, the fish trade had a wide range of trading connections by the late eighteenth century. As early as the 1770s inland markets were the most important outlets for the Yorkshire coast industry. North Sea fish was supplied by cart and pannier pony train to numerous towns and cities across the county. By the 1780s Manchester had a regular supply, delivered by pannier pony train, from the Yorkshire coast and some even found its way on to Liverpool. Much of this fish came from Staithes, then probably the largest fishing station on the English North Sea coast north of the Wash. Other consignments went up the Humber estuary by way of Kingston upon Hull, where they were loaded into river craft which took them up the Don as far as Sheffield or along the Trent to Gainsborough and beyond.[16]

North of the Border, Edinburgh was an important market for fresh white fish taken by small undecked line boats operating from villages and towns on the Fifeshire coast. Further north, a large quantity of fish was cured and by 1800 several villages had established strong reputations for smoking haddock, particularly Findon which gave its name to the process by which the fish, after splitting to present a flat surface, was smoked over peat fires.[17] In the English Midlands, Birmingham already had a well-established fish market and catches landed in the South West not only sold in regional centres such as Exeter, Bath and Bristol but as far away as London's Billingsgate, an ancient market which received supplies of fish by land as well as water. Parliament was ever anxious to improve London's fish supply and in 1761 it passed an Act granting fish vans freedom from post-horse duties and exemption from turnpike fees on their empty journeys home.[18]

Yet even though late eighteenth-century Britain possessed a fish trade of considerable sophistication, with a wide range of national and international connections, there was still room for massive expansion, as nineteenth-century developments illustrated. Although there is evidence to suggest some growth at the time, expansion was not dramatic when compared with the changes being wrought in agriculture, as village after village enclosed their open fields in response to the growing population's increasing demand for food.[19] The bottleneck which restricted a greater rate of expansion for the fish trade was the high cost and slow pace of contemporary transport which posed acute problems for a perishable commodity like fish. True, pannier ponies did transport fish far and wide but the operation, though quite swift, was expensive as only a relatively small load could be carried per animal. These costs usually limited inland trade in 'fresh' fish to prime varieties, sole, turbot, cod, ling and the like, destined for the wealthier end of the market. The cheaper 'offal' fish, including fresh haddock, plaice and coalfish, could not bear the cost of distant transport. Thus the inland poor could not usually afford to buy fresh fish; but in coastal districts, where it was often sold by fishwives and hawkers, it was a readily accepted part of the diet. It is sometimes suggested that the inland fish trade was limited because the poor considered fish to be an inferior food, but even though conservatism in diet may have been a factor, a great deal of contemporary evidence suggests that the problem was the type of fish they were offered. Much fish only came into their price range because its quality had deteriorated making it 'generally half rotten and consequently most unwholesome and disgusting food'.[20] Then, of course, there were herrings which were caught seasonally in such numbers that they were forwarded by the cart-load to towns. However, herring being an oily fish deteriorates particularly rapidly if left uncured so many consignments reaching eighteenth-century London must have begun to turn on arrival which can scarcely have made them the most attractive of foodstuffs.

Theoretically, curing could solve many problems but in practice the quality of the finished product often left much to be desired. The only way that fish could be kept for any length of time was by heavy curing. Until the nineteenth century, when government agencies intervened, British curing standards were notoriously variable. When fish were dried, for example, much care was needed to prevent them becoming burnt by the application of too much salt, or blistered by prolonged exposure to strong sunlight; if they were not stored in an absolutely dry place then mould would soon appear.[21] Quality was obviously appreciated, but a lot of cured fish was treated with reserve by all classes of British society and this retarded the growth in demand.

Obtaining supplies of fish for the growing towns and cities acquired a new importance in times of poor harvest and shortages of staple foodstuffs. Yet curers could not respond swiftly to the opportunities thus presented. Red herrings had to spend up to three weeks in the smokehouse and throughput could only be increased by taking short cuts which lowered quality. Furthermore, salt could not be obtained swiftly, for the only native supplies suitable for curing came from Cheshire.[22] This had to be forwarded to Liverpool and then sent coastwise, so much time could elapse between ordering and delivery. Moreover, the Salt Laws inhibited the curer from hoarding supplies of salt in readiness for any surge in demand. Under this legislation all salt was liable to duty and this was an extremely valuable source of state revenue. To ease the situation for fish curers the Government exempted them from the duty in 1786. To obtain exemption, however, they had to follow strict regulations involving the construction of costly, bonded warehouses, detailed accountability of all salt used and regular checks by Excise officers.[23] Such bureaucracy reduced the incentive to retain large supplies. The Salt Laws were not finally abolished until 1826.

These problems made it difficult for the eighteenth-century fish trade to develop a mass inland market amongst the growing urban working classes. There were many occasions when inland areas were suffering severe shortages of provisions whilst the coastal fishermen were catching fish in such quantity that they could not always be disposed of. Some towns attempted to offset the fish trade's transportation costs by offering financial incentives—known as bounties—to fishermen and merchants who would bring them supplies. Hull and Bristol both operated such schemes at various times and during the French Wars the Government embarked upon a national project of this nature. In 1800 an official report concluded that obtaining as much fish as possible from the seas was by no means the only object and that securing effective distribution was only possible with proper organization. It advocated the formation of local voluntary societies to organize the supplying and marketing of suitable fish. Such societies were envisaged for London and other ports at the mouths of inland waterway networks, including Bristol, Hull, Liverpool and Lynn.[24] Although the Government backed the scheme, and made available loans of up to £7,000 interest free to any newly formed society, providing it could match the sum borrowed, it seems to have met with little success for the bounty system was increasingly adopted as the provisioning crisis became acute. From September 1801 the Treasury was empowered to grant bounties for the bringing of fish to the cities of London, Westminster, Edinburgh, Exeter and other places.[25] Supplies to London and most of the other places involved seemed to have improved[26] as a result, but to what extent this increase in supply was matched by an increase in catching capacity is uncertain. The fishing fleet would seem to have been depleted by the depredations of enemy privateers and, more crucially, the labour force almost certainly shrank as many fishermen were pressed into naval service. Thus it would appear that the ability of the fish trade to maintain or improve overall catches was reduced. It is likely that much of the supply normally disposed of in the locality of many fishing stations was diverted to the bounty centres leaving such places as Scarborough without adequate supplies.[27]

It is apparent from this brief survey that the fishing industry in the pre-trawling era was both complex and considerable. Yet it would clearly have played a much greater role in provisioning the kingdom if large consignments of fresh fish could have been moved more cheaply and swiftly to the growing urban centres. Apart from the problems with curing, there were limits on the fresh fish market inland as only prime varieties could stand the cost of distant transportation and this was reflected in the modes of capture deployed. Trawling found little favour in many coastal districts because the trawl net caught large quantities of less saleable fish—the so-called offal varieties. On the other hand, long lines and hand lines were popular with fishermen because they mainly took the largest and most sought-after fish—the prime varieties. Contemporary economics dictated that the hook rather than the net was the best way to catch white fish.

2

The Pioneers

The modern motor trawler is outwardly a highly refined fish catching machine. Its bridge bristles with sophisticated electronic equipment used for navigating the seas and locating an underwater quarry which is then taken in nets made from synthetic fibre. Yet despite such technological appendages it remains at heart a hunting craft, for the principal techniques of trawling have changed little over the centuries.

The trawl is still basically a bag-shaped net. Its fish catching efficiency is largely determined by the width of its mouth and before the 1890s this was usually kept open by a large beam or length of wood; hence the name beam trawl. The beam—often made from two trunks of ash, beech or elm scarfed together in the middle by iron hoops—was raised up some three foot on top of two stirrup-shaped iron supports or 'heads' which skidded along the sea bed. The main body of the net consisted of four sections, back, belly and two wings. The back or upper part of the net was fastened to the beam whilst the wings, which helped give the net shape, were secured on the trawl heads. The belly dragged along the bottom and was finished off at the forward end with a ground rope often fitted with small bobbins or rollers to stir up the fish. This net, made from tarred hemp, was in later days frequently over fifty feet wide and upwards of one hundred feet in length. It tapered away to a narrow trap, often fitted with pockets and known as the cod end. The trawl was dragged behind the boat by a single six-inch diameter warp of about 120 fathoms in length. This divided into two bridles which were attached to the trawl heads. Originally, the gear was hauled over the bow but later Barking fishermen

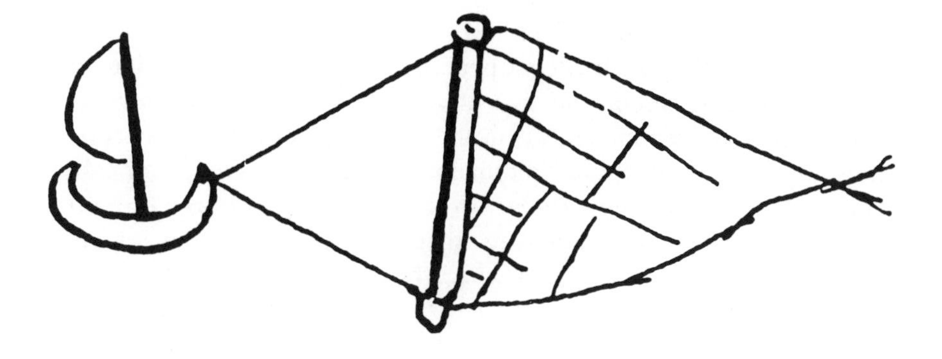

Fig. 1. 'The Trawle'. This drawing on the back of a state paper of 1635 is obviously of a beam trawl

introduced a capstan and after a six-hour trawl the net was then hauled in over the side; a back-breaking procedure which took the crew of five between two and three hours, sometimes in appalling conditions.

The earliest English description is of a device known as a wondyrchoun and appears in a petition from Thames ports to Edward III in 1376 against its continued usage. The petitioners likened it to a large oyster dredge with a close-meshed net from which no fish could escape and it was said to run so heavily over the sea bed that it destroyed plants and the spat of oysters, mussels and other fish. So much fish was claimed to be taken by the wondyrchoun that fishermen could not sell them all and used the residue for fattening their pigs. A commission set up in response reported the following year that the 'machine' was three fathoms long with a beam of ten feet, at the end of which were two frames shaped like 'cole rakes'. In short, it was a form of beam trawl. The commissioners recommended that the net be used only in deeper offshore waters but no legislation was passed on the matter and the practice was left to develop.[1]

Trawling soon came to be despised by other fishermen and eventually attracted the attention of the legislators. In 1491 many trawlermen working on the Norfolk and Suffolk coasts were fined the then considerable sum of £10 for fishing with small-meshed nets and 'unlawful engines' for taking small fish.[2] In 1622 the Mayor of Hythe complained to Lord Zouche, Warden of the Cinque Ports, about fisherman from Rochester and Stroud who were allegedly trawling with illegal nets. Charles I received petitions in 1630 protesting about the size of meshes used by some trawlers. Throughout the seventeenth century strict regulations were enacted prohibiting the use of small-meshed nets and these were sporadically enforced with the utmost rigour. In the first year of George I's reign a further Act was passed stating 'that any person using at sea any traul net, drag net or seine net which has mesh less than 3½ inches from knot to knot shall forfeit such nets'. The offender faced a fine of up to £20 and the nets were to be burnt.[3] This legislation was still being enforced over 120 years later.

The activity which was the object of this legislative attention was at the time a relatively unimportant means of taking fish and largely confined to inshore waters. But its use was not limited to British coasts. Indeed, the trawl is said to have been derived from very basic types of drag net used on various European coasts; these included the Mediterranean *aissaugue*, the *chalut* of Poitou and the *gangui* of Languedoc. Manorial records show that a horse-drawn trawl had been used on the shores of Filey Bay[4] and other places for centuries and a type of primitive bag-like trawl was said to be still in use on the coast of Brittany in the early part of this century.[5] Beam trawls were also used in Dutch waters from at least the middle of the seventeenth century and proved almost equally controversial.[6]

By the latter half of the eighteenth century the English trawl fisheries were largely concentrated on Devon coastal waters and the approaches to the Thames which were worked by craft hailing from Brixham and Barking respectively. Though the early Torbay trawlers probably worked coastal waters they gradually moved on to deeper-water grounds.[7] Yet trawls were being deployed in other districts: for example small trawlers were working out of Ramsgate as early as 1790 for their crews are recorded as claiming freedom from impressment, while a small-beam trawl was being used by Hartlepool fishermen soon afterwards.[8] In 1770 an unsuccessful attempt had been made by the Welsh to emulate Brixham as a trawling centre when a fishery society established at Swansea brought across an old Torbay fisherman and his boat.[9]

The French Wars seem to have acted as a stimulus to this

embryonic industry encouraging at least a temporary movement further afield, particularly in the Southern Bight of the North Sea. During the later 1790s some forty sail belonging to Barking found employment trawling on grounds such as the Broad Fourteens and Brown Bent off Yarmouth. They were even recorded as far north as Smith's Knoll and apparently worked almost across the North Sea to Holland.[10] Devon fishermen were to be found in the English Channel off Dover as well as the Bristol Channel. Walter Smith, a Brixham fisherman, told a Government inquiry in 1837 that he had originally fished off the Kent coast some forty-five years previously and claimed to have been the first from Torbay to go fishing so far eastward.[11]

As problems of contemporary transportation restricted the distant movement of fresh fish to those more valuable species that could stand the cost of conveyance in relatively small quantities, it is at first sight surprising that the practice of trawling—which took such a large proportion of less valuable fish—should be on the increase in the later eighteenth century. However, a closer examination reveals several factors which encouraged this modest expansion. The nation's population was growing rapidly and more people were moving from the country to the town. As towns grew in size it became increasingly necessary for the authorities to stimulate the supply of provisions including fish, particularly at times of acute shortage. Despite the difficulties associated with the movement of fish such factors could not fail to have a beneficial effect on some branches of the fish trade, including trawling. For the Barking trawlermen at least, the transportation problem was less important. They fished in the Thames approaches and, if necessary, could sail upriver to unload at Billingsgate itself, thus taking advantage of the country's largest market.

Though geography denied the Devonshire trade the convenience of an adjacent capital city, it did confer one advantage: no centre of population in that county—then the third most populous in Britain—was more than thirty miles from the coast. Many of the cheaper varieties of fish could therefore find markets nearby. At the same time the improving condition of roads in the turnpike era made it commercially worthwhile to forward the prime portion of the trawler's catch further than just Exeter, Bristol or even Bath. London had an insatiable appetite for quality flat fish and the

Brixham firm of Grant and Company forwarded over £2,800 worth of sole and turbot there between 1765 and 1767 alone.[12] Fishing had always been an important part of the Devon economy and over the latter part of the eighteenth century trawling became an increasingly significant element. Brixham, of course, was at its centre and could boast a fleet of over a hundred trawlers as early as 1786. The other Devon trawling centre was Plymouth but there the pursuit of herrings and pilchards were far more important activities and even as late as 1820 the port could claim no more than about thirty trawlers.

The early eighteenth-century Torbay trawling vessels were quite small but after about 1750 they gradually increased in size and became completely decked. By the mid-1780s they had evolved into carvel-built, beamy vessels with straight stems and deep counter-sterns, a basic hull form they were to retain until the end of sail. Initially, these craft were sometimes described as sloops or, more fully, as cutters but later were better known as smacks. As the hull design improved so too did the rigging. They were single-masted with a gaff-rigged mainsail, a bowsprit and a jib mainsail.[13] They were extremely seaworthy and ideally suited for working on the offshore grounds being increasingly opened up. Even so, they were still somewhat smaller than the deep-water craft used in other branches of the fishery. The typical 1790s larger class of Torbay smack was probably no more than thirty-five feet from stem to stern whilst the three-masted luggers found on the Yorkshire coast and other places measured up to fifty-four feet in length[14] and even these craft were small in comparison with the Dutch herring busses. Moreover, a trawler rarely carried above five crew whilst the large herring craft generally shipped a minimum of eight.

The provisioning problems, which had encouraged some spread of trawling during the French Wars, ended with the exile of Napoleon to St Helena. A run of good harvests and the availability of corn imports from Europe brought down food prices which reduced agricultural prosperity and led to protectionist measures in the form of the Corn Laws of 1815 aimed at stemming the flow of imports. Despite such Government intervention, agriculture entered a depression which lasted around thirty years.

Not surprisingly, the price of fish—which had long enjoyed a degree of protection from foreign imports—also fell and was generally to remain low on the Billingsgate Market until the 1830s.[15] Lower prices affected the Devon trawling trade in particular, given the high costs it had to sustain for transporting fish to distant markets from Brixham and Plymouth. The Torbay fishermen, however, responded positively to the new environment and began working further from home. As they moved eastwards along the Channel they made seasonal bases as far away as Ramsgate and Folkestone. This movement enabled them to supply the London market with quality fish, including Dover Sole, more cheaply by shortening the costly overland journey to London. At Ramsgate a fleet of up to seventy Brixham smacks would arrive in November by the later 1820s. Some would merely stop until Christmas whilst others remained until the middle or end of June, after which they returned home because the weather became so warm that they could not send their fish into London fresh enough for the market.[16] Their arrival supplemented the activities of the inshore trawlers in places like Rye Bay but these smacks when sailing out of Ramsgate did not confine their activities to the English Channel. By about 1825 they too worked into the North Sea as far as the East Anglian and Dutch coasts. On venturing into such strange waters they often sought the assistance of the Barking men who—having considerable knowledge of suitable trawling grounds—were often recruited as pilots. By the 1830s both Ramsgate and Dover had become the home base of many of these smacks.

The opportunities presented by the markets of Bristol and Bath encouraged a similar interest in the Bristol Channel whilst Brixham trawlers were soon attracted to Dublin and Liverpool. By the 1830s vessels from the latter port had already adopted an early version of the later fleeting system, for at certain seasons of the year they would rendezvous at Douglas, Isle of Man, in order to work a bank off the island's eastern coast, and send their fish in by the steamers which were now making their mark in coastal waters. Sometimes, when the weather was favourable, one of the smacks would carry in the fleet's catch and be entitled to a share of the total cargo as a reward.[17]

The activities of the Barking men in the North Sea had also generally increased since the end of the Napoleonic Wars though few of their craft specialized in the practice: most of them trawled only in the winter and went lining for cod in other seasons. The trade was organized differently on the Thames where ownership of the fleet was largely concentrated in the hands of a few individuals or firms—the most notable of which was Hewett's—whilst there was a strong tradition of single-boat ownership amongst the Torbay men.[18]

However, the Barking fishermen confined their trawling activities largely to the Southern Bight of the North Sea and it was the Torbay smacks which were to take the lead in spreading the activity further up the east coast. Their initial involvement with the Yorkshire coast, however, was at the behest of Hull Corporation and Colonel Ralph Creyke of Marton Hall near Flamborough. The first experiment in offshore trawling on the Yorkshire coast seems to have taken place in 1819 when a Flamborough vessel, registered at Bridlington, worked a fishing ground lying south-east from Dimlington Heights in south Holderness. The initiative appears to have met with some success for more than a hundred pairs of sole were taken with one trawl of the net. Creyke, a local magistrate and treasurer of the Flamborough Fishermen's Fund—a friendly society set up in 1809—was probably behind this experiment. In June 1819 he had purchased a thirty-nine foot, thirty-three ton cutter-rigged craft, *Moor Park*, which was operated for him by Cornelius Young until the following October. In 1821 the Bench of Hull Corporation sent Creyke to Plymouth to tempt trawlermen to try their luck out of Hull.

The inducements offered were fairly lucrative and John Davis of Plymouth, master of a thirty-three ton craft, eventually decided to venture north. Under the terms agreed he was to receive one guinea from the town clerk on his arrival in Hull and a further twenty guineas, spread over three months, for landing at the port. The Corporation also agreed to pay an able pilot to sail with him for the first month. In the event two smacks turned up and Creyke encouraged the Corporation to offer the same terms to the second skipper. Some arrangement was certainly made because both smacks stayed for some time. The venture was a limited success. Both skippers became convinced of the potential for trawling on the Dimlington grounds but were dogged by their unfamiliarity with the sea bed. Their beam trawls could only operate upon smooth-bottomed grounds and they were unfortunate in fouling their nets and badly damaging them on underwater obstructions. Certainly the experiment was not persevered with in the following years but some idea of the area's potential had been gleaned by the south-western fishermen.[19]

Some ten years later, in 1831, the trawlermen returned. That summer two smacks were working on Yorkshire coast grounds and landing their catches at Scarborough. These visitors found a number of attractive commercial features which encouraged them to stay. The town was a fashionable resort for the wealthy and their household entourages, and their influx each summer considerably increased the local population and thus demand for provisions. Yet this sharp upturn in demand occurred at precisely the time that local fishermen were busy supplying other outlets, both inland and overseas, so the smacks were able to exploit this gap in the market and so establish a pattern of seasonal landings.

Sometimes, the sheer quantity of fish landed by just these two trawlers had a detrimental effect on prices. Such occurrences inflamed local fishermen and periodic disturbances erupted. In July 1831 the local men asked Scarborough magistrates to prevent the two West Country smacks coming in to sell their fish. On being told by the magistrates that they had no legal right to interfere, men from Scarborough—and probably the surrounding fishing communities—decided on direct action and the following day a large crowd gathered on the beach by the fish market intent on preventing the smacks from landing. This action proved only a partial success for, although the two boats avoided confrontation by making off, one was to return when the crowd had dispersed and successfully landed its catch.[20]

These smackmen must have found their visit worthwhile because the following year they returned in greater force. In late May 1832 some eight smacks, hailing from Ramsgate, Dover and Plymouth, arrived at Scarborough. They obviously intended to stay for some months as the crews brought their families as well. Their arrival aroused the previous year's animosity and in early June there was a series of brawls and a southerner was stabbed by a local. To restore order the local magistrates had to swear in the area's preventive men as special constables.[21] After this an uneasy peace was secured.

The southerners continued to visit throughout the 1830s, and though there were occasional outbreaks of unrest the remainder of the decade proved prosperous for all and this helped to alleviate the situation. Vessels also began landing at Hull to take direct advantage of its market and the network of inland trading connections—along which fish already moved—provided by the Humber and its tributaries. In 1834 Thomas Sudds, late of Ramsgate, moved permanently to the port and registered his craft there. Yet, at both Hull and Scarborough the activities of the southerners remained predominantly seasonal throughout the remainder of the 1830s and only a handful of other smacks settled permanently. The first two Scarborough smacks were *Forager* and *Providence* which were registered in 1839 and 1840 respectively. *Forager* was skippered by Thomas Halfyard, who hailed from Ramsgate and who later figured prominently in developments at Hull.

Thus, between the 1790s and the end of the 1830s trawling had begun to spread from its Devonshire and Thames roots along the coasts of England and Wales and out into deeper waters. Though new trawling bases were established at, for example, Ramsgate, Liverpool, and Dublin, trawling formed a minor part of the national fish trade and many smackmen remained only seasonal visitors to the new grounds they were opening up.

3

Railways and Markets

During the four decades following 1840 the British fish trade underwent several changes which, taken together, were little short of revolutionary. Firstly, there was an almost nationwide upsurge of activity and output, if the report of the 1866 Royal Commission on Sea Fisheries is to be believed. Secondly, most of the North Sea grounds were trawled for the first time and trawling came to be carried out on such a scale that it rapidly surpassed lining as the principal means of taking white fish. Lastly, these years witnessed the establishment and rapid rise of the 'new' fishing ports of Hull, Grimsby and later North Shields where, because of the sheer scale of activity and the methods adopted of organizing capital and labour, the trade assumed some features of large-scale Victorian industrialism.

It is perhaps not surprising that these developments coincided with the construction of the national rail network. One obvious advantage of railways over earlier forms of transport was their speed. Henceforward, foodstuffs could be conveyed from port to market far more rapidly than had hitherto been possible. This could only be beneficial for the marketing and distribution of fresh fish and finally provided the means of breaking the transport bottleneck that had long constrained the trade. Ultimately, railways enabled fish to become an article of cheap mass consumption but, despite the growth that occurred in just a few decades, such potential was not realized overnight. Indeed, historians have sometimes expressed surprise that the railways failed to take over the movement of such perishable commodities more rapidly.

Few early railways were built with the idea that fish traffic would provide a considerable source of revenue. In the case of the horse-drawn Whitby to Pickering line, built between 1832 and 1836, even George Stephenson makes this clear. Although he wrote to the promoters in 1832 about fish, along with agricultural produce, timber, and whinstone as possible sources of income, he set much greater value on coal and lime traffic.[1] The promoters of the Hull & Selby line, which established rail links with the West Riding and Lancashire from 1840 also failed to recognize the potential of fish traffic when estimating the line's earnings. Yet this railway offered a fast and apparently convenient means of conveying fish inland. Fishing vessels from the Yorkshire coast as well as the southern smacks had established the custom of making seasonal landings at Hull, and activity had been increasing in the years prior to the line being built. Once opened, the logical step was to forward fish along it, but during the first eighteen months of operations the level of such traffic remained insignificant with no more than three-and-a-half tons of fish reaching the Leeds to Manchester line from this source. It appears that the railways had done little more than take over part of the traditional boat, cart and pannier pony traffic flowing along their routes.

This initial lack of awareness, or even interest, in developing the fish trade's potential seems to have been part of a wider attitude to new lines of business. Though some railways, particularly those on the north-east coalfields, created new traffic, most merely aimed to secure existing trade. It appears therefore that although the railways were initially content to take over the existing fish traffic, they appeared unlikely to stimulate further developments in the marketing and distribution of fish. It was observed in the early 1840s that the east-coast fisheries were making large catches but were unable to dispose of all fish landed even though there was widespread distress and want of cheap provisions in Manchester through the onset of the sharp trade depression of 1842. There, the price of fish remained high and cod fetched from 8d to 1s per pound, rarely falling as low as 4d per pound.[2]

This represented a classic example of the gap that separated fishermen from the mass of inland consumers. There had been many occasions when great need for cheap sustenance had been left unsatisfied even though large catches were being made on the coast. During the winter of 1766/7, for example, Scarborough fishermen located and caught huge shoals of haddock. Large quantities were landed and on many days the local market was quite glutted with small haddocks being sold off to the poor for as little as 1/2d per score. Sometimes supply completely overwhelmed demand and the fishermen had to lay up their cobles. Tragically, this local surfeit had occurred at the same time as severe shortages, caused primarily by poor harvests, had afflicted much of the rest of the country.

The limitations of conventional transportation in the pre-railway era meant, of course, that it was expensive to move perishable commodities such as fresh fish swiftly over long distances and this restricted the trade to the quality end of the inland market. Moreover, the nation's traditional transport infrastructure, particularly away from the coasts and waterways, found it notoriously difficult to respond to sudden demands brought about by, for example, the need to alleviate shortages: a situation all too common in Third World countries today. That this state of affairs should continue even after the railway network began to spread across the country was partly due to the continuing high cost of transport. The railways at first followed the example of existing carriers and, 'deeming fish a luxury which had to be conveyed', charged high carriage rates.[3] In many respects this reflected the general aims and practices of the early railway companies. The trunk lines were established by concentrating largely on high tariff business—quality rather than quantity. Apart from luring high-fare-paying passengers from the stage coaches, they also tried cutting high-value merchandise rates just enough to take much of it from rival road carriers. They were not at first interested in encouraging the development of lower-value traffic, including the cheaper varieties of fish, by cutting rates drastically, and failed to appreciate the radical potential their system of distribution could have on the market for such a perishable commodity.[4]

There was another aspect. The trade suffered, like others, from the parochial policies adopted by many railway companies to the movement of through traffic. Bagwell has pointed out that physical contact between railways was one thing but businesslike conduct of through-traffic arrangements was quite another.[5] Consequently the fish trade sometimes had to deal with the need for separate payments and transhipment at the boundary of each company's lines. Early carriage rates also took little or no account of different varieties of fish or the best conditions under which each might be forwarded. Such a combination of problems meant that at first it was usually worthwhile only to forward prime varieties, such as large cod, sole and turbot. Cheaper offal fish, including haddock and plaice, could not stand the cost of conveyance, so the railways, whilst speeding up deliveries, often did little to widen the market. What mass consumption of fish there was in inland northern industrial districts, for example, was usually limited to the heavily cured fish that could be forwarded by the cheapest and slowest means. As late as 1841 it was noted that fresh fish was rarely eaten by the Manchester working classes and consumption of any type was not common unless they were Catholic.[6]

The problems associated with travelling over several companies' lines brought complaints from passenger and freight customers alike and eventually prompted railway companies to try and harmonize their working arrangements. In the spring of 1841 the boards of the Manchester & Leeds, Leeds & Selby and Hull & Selby companies reached agreement on a scheme to encourage through-traffic. All agreed that receipts should be divided in proportion to the route mileage of each concern the traffic travelled over. Even before this, one individual at least had recognized the potential advantages of railways for the fish trade. In 1839 Christopher Tennent of Hartlepool, a key figure in the emergence of the town as a modern port, aimed to use railways to make it a major fishing station. Unfortunately, Tennent died the same year whilst visiting Leeds for this purpose and much of the initiative's energy passed with him.

However, in the latter half of 1841 Captain James Laws RN, manager of the Manchester & Leeds Railway, embarked upon a scheme to encourage fish traffic by adopting more sympathetic carriage rates. After negotiating with the managements of the two other companies on the line to Hull, and travelling to meet Hull, Filey and Flamborough fishermen, he was able to secure the introduction of a more attractive and cheaper range of carriage rates; one shilling per hundredweight for the journey to Manchester for mixed baskets of fish, providing the fishermen undertook to sell them for no more than two shillings per stone. Captain Laws also ensured that an area was set aside in Manchester on the Salford side of Victoria Bridge, where a number of people from the fishing communities involved opened a shop cum stall in the name of the Flamborough and Filey Bay Fishing Company. To ensure supplies, a tight schedule had to be adhered to, particularly by the Flamborough and Filey fishermen who, because the railways had not reached them, had to carry the fish by cart overnight to Hull in order to catch the 6 am train. If this was achieved, the fish arrived in Manchester for noon and could be on sale within twenty-four hours of being landed.[7]

The shop opened at the end of January 1842 and received fresh supplies six days a week. It proved an immediate success. There was an overnight drop in the price of fresh fish from between 8d and 1s per pound to 1¾d per pound, and the poorer classes flocked to buy in such numbers that for a time the footpath over the neighbouring bridge was completely blocked. The entire consignment of 3,192 lb was disposed of in under two hours. The story was much the same over the following weeks and fresh fish was soon established in Manchester as an article of cheap mass consumption. A few years later it was stated in the House of Commons that Laws' arrangement had 'brought the commodity within the means and inclination of so large a class of people as to raise a demand that has kept ahead of supply . . . and it has led to the habitual use of fish by a large number of persons who rarely tasted it before'. Within three years the amount of fish handled by the Manchester & Leeds line had risen from 3½ tons to 80 tons per week, proving that the demand for fresh fish amongst all classes was there at the right price and quality. The experiment was not limited to Manchester for the new rates applied to all stations along the route, and by 1845 a great deal of fish arriving in the city was being redistributed to adjacent districts. Supplies were gradually drawn in from other stations as the pace of railway construction increased during the two railway manias of the 1840s.[8]

Despite being a major step forward, Laws' arrangement had its limitations. The markets that opened up, for example, were restricted to those towns on the lines which had organized the scheme. Moreover, it did not provide the desired range of rates for the different varieties of fish, and the imposed maximum of 2 shillings per stone sale price could represent a price ceiling at times of intense demand. Yet the basis of the agreement—without the ceiling—seems to have survived the takeover of the Hull & Selby concern by the York & North Midland Railway in 1846. This in its turn became a principal constituent of the North Eastern Railway in 1854.

Under the original agreement the companies had divided traffic between Hull and Manchester on a mileage basis and this principle was maintained when the Railway Clearing House, created to develop a systematic means of managing through traffic, was founded in 1842. The companies operating between Manchester and Hull were joined by five other concerns including the London & Birmingham and York & North Midland railways. Gradually, most companies were to join and as amalgamations strengthened the hold of key members the Clearing House was able to devise or influence traffic arrangements across the country.

The Clearing House was faced with a difficult task and it is perhaps not surprising that its monthly meetings were at first concerned mainly with passenger fares and wagon rates. It was not until 1847 that it began to tackle the formidable diversity of practices affecting goods traffic. Even then, it was November 1849 before its attention suddenly turned to developing a range of rates for fish and this was partly due to lobbying from the fish trade. It was also due in part to the changing economic environment. The marked extension of the network during the 1840s intensified competition for traffic and affected profitability. Railway companies were forced to look for new areas of business. Less emphasis was given to quality traffic and more to quantity. With passengers this led to a shift of emphasis from first to third class. In the case of the fish trade, steps were taken during the 1850s which assisted the development of rail traffic in lower value fresh fish that could enjoy mass demand.

However, the process of developing a complete and cohesive rating policy for fresh fish from all fishing ports proved a tortuous business, requiring several bones of contention to be settled. Firstly, the Clearing House had to determine whether fish should be defined as goods traffic and assigned to the goods ledger or, bearing in mind that much went by passenger train, as parcels traffic and entered in the passenger ledger.[9] Such an issue might seem trivial but the development of through-traffic depended on the matching up of accountancy practices. Then there was disagreement over the terminal rate. This was the amount of money, in addition to the mileage share, that went to companies loading and unloading fish at either end of the journey. The final and most divisive issue concerned risk. The trade wanted the option of sending fish at the carriers' risk whilst the railway companies were keen that such perishable consignments should be forwarded at the sender's risk. The issue was a constant source of friction which resulted in law suits and thus delayed resolution.[10] Matters came to a head for the railway companies in 1855 when the Manchester, Sheffield & Lincolnshire Railway, intent on developing Grimsby, offered the fish trade there the most reasonable rates. This threat of competition produced results and early in 1857 a range of comprehensive rates for most varieties of fish from the various ports was drawn up. The railway companies conceded the carrying of fish at their risk but offered a cheaper alternative range of prices for senders who accepted the risk themselves.

The formulation of nationally agreed procedures for dealing with fish traffic and the continued extension of the railway network was reflected in a threefold increase in the number of English fishmongers between 1841 and 1871 and a widening range of market penetration, as can be seen from a study of the Yorkshire coast. During the 1840s, for example, only a small quantity of the fresh fish landed there was distributed further than Lancashire. Though Billingsgate had received its first rail-borne supplies about 1846 it is clear that the only fresh fish forwarded there from Yorkshire even as late as 1854 was high quality salmon; yet by 1863 the Yorkshire ports, together with Grimsby, had become major suppliers of fresh fish to London.[11]

Henceforward, railways nationally provided a cheaper as well as faster means of transport. Price reductions on inland markets were largely due to a lowering of carriage costs. At the same time fishermen and merchants all round the English coast reported that landing prices remained bouyant despite a rapid expansion in the amount of fish being landed. By 1866 the range of places supplied with white fish by Hull, Grimsby and Yarmouth was much as it was to be for the following hundred years.[12] In the South West, Brixham amongst other places noted a great upsurge in the profitability of trawling in the later 1850s. In the North Sea, many cheaper grades of fish, caught in great numbers by trawlers and often previously thrown back as unsaleable could in future bear the cost of forwarding to inland markets. Hull fishermen reported that their smacks threw little back after 1857.[13]

In order to develop the new mass markets a wider range of commercial outlets had to be opened up. Both the North Eastern Railway and the Manchester, Sheffield & Lincolnshire Railway provided free travel for Hull and Grimsby fish merchants when they journeyed to inland towns and cities to establish new contacts. The creation of the national telegraphic network during the 1840s and 1850s also helped nurture the national market, for it enabled both inland buyers and salesmen on the fish quays to ascertain where the best prices, supply or demand lay. For the majority of the labouring classes crammed into overcrowded tenements in the teeming industrial towns of the English heartlands, really fresh fish was a novel addition to the diet. In London the railways finally made fresh fish widely available to the poor throughout the year rather than on a seasonal basis. However, the arrival of cheap fish in these areas also coincided with an influx of consumers for whom fish was already an essential part of their diet. They were the Catholic Irish and the demand they created at this crucial period in the development of the national fish trade should not be underrated.

Irish immigration into Britain had been running at a high level even before the Potato Famine of 1845, thanks mainly to population pressure on a backward agricultural economy. According to the Census there were 415,000 Irish born persons living in Britain by 1841. The failure of the potato crop on which so many depended led to hunger and death on an unprecedented scale and turned Irish emigration into a torrent which lasted at least twenty years. It has been estimated that 374,000 Irish emigrated to Britain in the decade 1841–1851 alone. According to the 1861 Census there were some 806,000 Irish born people in Britain, comprising some three per cent of the population, without taking account of their offspring born after arrival. In some areas, however, the density of immigrants was far greater: in Lancashire, for example, they made up nine per cent of the population.[14] The bulk of these newcomers were of Catholic peasant stock, whose religious beliefs dictated that fish was traditionally consumed every Friday. They flocked to the new industrial centres at the very time that railways were beginning to provide mass transport for the cheaper varieties of fish, a source of demand that must have been a powerful additional stimulus for the new northern trawling ports of Hull and Grimsby in particular.

Moreover, it also seems that whatever reservations the British working classes may have held in the pre-railway era, those in the burgeoning industrial centres gained a considerable appetite for the cheaper and fresher fish they now received by rail. This was soon to be reflected in the rise of that British institution, the fish and chip shop. However, the fried-fish trade had a somewhat inauspicious beginning—at least in London. Dickens's *Oliver Twist*, written in 1850, has a brief reference to a fried-fish warehouse in Field Lane, next door to a beer shop. By the time Henry Mayhew wrote *London Life and Labour* in 1861, fish frying was becoming an established part of the life in the capital's teeming slums. He estimated that there were approximately 300 fried-fish sellers, most of whom were concentrated within three miles of Billingsgate. Fried fish was then usually sold from stalls or hawked by itinerant sellers—costermongers—in neighbouring streets and pubs. The fish they used was known in the trade as 'friers' and comprised the stock that the fishmongers had not sold overnight and did not wish to offer again the next morning. Instead, they sold it cheaply to the friers who washed and, if necessary, gutted it before cutting it into pieces, dipping it in flour and water and frying it in an ordinary frying pan. These early fried-fish sellers lived mainly in the labyrinth of courts and alleys running from Gray's Inn Lane to Leather Lane, between Fetter Lane and Chancery Lane, as well as in courts around Cowcross Street, Smithfield and Clerkenwell. They were also to be found in the alleys about Bishopgate Street, Kingland Court and in half-ruined rookeries near Southwark and Borough Roads. One costermonger told Mayhew that a gin-drinking neighbourhood suited best 'for people haven't their smell so correct here'.[15]

The itinerant sellers can hardly have sold their wares in the hot and fresh condition we are accustomed to today. Nevertheless, they satisfied a demand and found setting up in business to be a comparatively cheap affair. These street traders usually possessed a neatly painted wooden tray, often covered with fresh newspapers and strung by a leather strap from the neck. Upon this they spread the shapeless brown lumps of fish. Parsley was strewn over them and a salt box was provided for the customer's use. Each tray generally contained from two to five dozen pieces of fish. Equipment and stock cost no more than ten shillings and less if second-hand gear could be obtained. Stallholders, of course, required more capital.

Sometimes these traders ventured further afield, for Mayhew noted that sales of fried fish had become a feature of neighbouring fairs and races. At Epsom Races the fried-fish sellers found customers amongst those who worked and served around the meeting. Bread or relish made it into a meal and plenty could be picked from the ground where it lay after being flung in great amounts from the carriages of the well-to-do after lunch. Bread was also a common accompaniment in London at this time, as were baked potatoes. The link between fish and that French invention, fried potatoes, which was to revolutionize the trade, was not made until after the 1860s. It is unclear when fish and chips were first sold together.

By 1876 the trade was established in shops but these early establishments left much to be desired. The pans consisted of open cauldrons built into brickwork and fired by coal. It was noted that friers were often ex-stokers and conditions may have seemed more akin to a boiler room than a modern fish shop. An account of one such early frier describes how, between serving, 'firing up' and 'stirring up' the contents of the pan—operations which were carried out with as much splash and splutter as possible—he would stand behind the counter chewing away at a great quid of tobacco with his greasy hands in his pocket. Conditions were often unhygienic, with the offensive smell of boiling oil and frying fish permeating the entire street on which a shop was situated. The nuisance they caused regularly occupied health officers across London. Conditions improved all round after Lancashire engineers developed improved ranges which carried away the steam and smoke and most of the abnoxious smell. Lancashire soon became the centre of this trade and indeed John Rouse (Oldham) Ltd claims to be the firm which popularized the sale of fish and chips. It has been suggested that the high proportion of women and girls employed in the textile industry on both sides of the Pennines originally encouraged the spread of the fish and chip shop because it provided a quick ready-made meal for their families. Fried fish and chip shops and the standard of their product continued to improve and ultimately they became a respectable, and traditional, part of British life. By 1913 it was estimated that there were over 25,000 fried-fish shops across the country consuming at least a quarter of Britain's annual catch of 800,000 tons of fish.[16]

Thus the railways eventually created the means of opening up a whole new range of marketing opportunities for fresh fish by providing the first cheap and swift mode of conveyance from coast to consumer. It is true that some growth would have occurred without them: on the supply side improvements continued to be made for many decades in the transport of fish by sea, especially to London. On the demand side, railways became readily available to the fish trade at a time of particularly rapid migration of people—both British and Irish—from rural areas to the growing industrial towns and cities: people who required a cheap and nutritious form of sustenance. Without the service provided by the railways, fresh fish would simply not have become such an important part of the national working-class diet in the late nineteenth century.

4

Opening up the North Sea

The creation of a national market for fish by means of the railways was, of course, accompanied by massive growth in the catching sector. Most types of fishing activity, whether inshore or offshore, for herring or white fish, seem to have been revitalized by the expansion of marketing opportunities; but none performed better at this time than the English trawling trade. Between the 1840s and mid-1860s there was a dramatic increase in the levels of trawling activity off the English and Welsh coasts and it rapidly overhauled lining as the principal means of taking white fish. Furthermore, the dramatic rise to prominence of the new fishing ports of Hull and Grimsby was based mainly on fishing with trawl nets.

Three factors generally feature in most explanations of why the trawling trade, particularly in the North Sea, did so well at this time. Firstly, it has been suggested that the rise of Hull and Grimsby and the opening up of the North Sea grounds was due to the exhaustion of the south-western trawling grounds which caused the migration of the Devon smacks. This thesis can be discounted, for though there was certainly a major migration of Devon smacks around mid-century, this coincided with a period of increased investment and sustained growth at both Brixham and Plymouth on a scale inconsistent with the picture of an exhausted and declining fishery.[1]

Great emphasis has often been placed on the discovery of the Silver Pits fishing grounds. Legend has it that their discovery was a pure accident. One year a few of the smacks which fished seasonally in the area were tempted to stay later than usual by a keen demand for their catches. As weather conditions worsened and severe cold set in they were inevitably caught by a tremendous gale and one smack, unable to retrieve its trawl in the storm, was blown far from the familiar grounds and across an area of deeper water. When the storm abated and the badly damaged trawl was hauled aboard its remnants were packed with the finest soles ever seen. Afterwards the battered craft made Scarborough harbour and then with the aid of a chart its crew retraced their route, eventually discovering the Silver Pits—the richest winter sole grounds in the North Sea.

There has been much uncertainty about when this discovery was made and the Silver Pits first exploited. Several dates have been put forward between 1837 and 1850, but part of this confusion may have been caused by contemporary sources referring to different grounds. The Little Silver Pit is about thirty miles to the east of Spurn Point while the Great Silver Pit is further to the north-east and closer to the Dogger Bank. These were just two of a whole range of grounds worked by trawlers from this time forwards. Others included Botney Gut, the Hospital Ground and California. It is not really accurate to say such grounds were discovered by trawlermen. A number had long been worked by local fishermen, especially the ground south-west of Flamborough Head which the trawlermen came to call California due to the prosperity it brought them. The Silver Pits were of particular importance because of the large hauls of soles they yielded and it was in this sense that the trawlers discovered their value. The local men had mainly deployed great lines which did not take many such flat fish. Moreover, the best sole season in the Silver Pits was winter when the large local fishing luggers were traditionally laid up.[2] So it was left to the trawlermen to realize their potential.

According to some contemporary sources it was William Sudds who 'discovered' the Silver Pits and research has shown he was one of the very first trawlermen based in the area. In 1833 he registered his Margate-built smack *Betsy* at Hull Custom House and the craft remained on the Hull register until June 1838 when it moved back to Ramsgate. William Sudds was certainly back in Hull again no later than 1844 when he was master of an 18-ton smack *Ranger*, the property of J. Todd, a local fishmonger.[3] It has been suggested that the Silver Pits were located in 1838, lost, then rediscovered in 1844,[4] which would in part account for the differing dates put

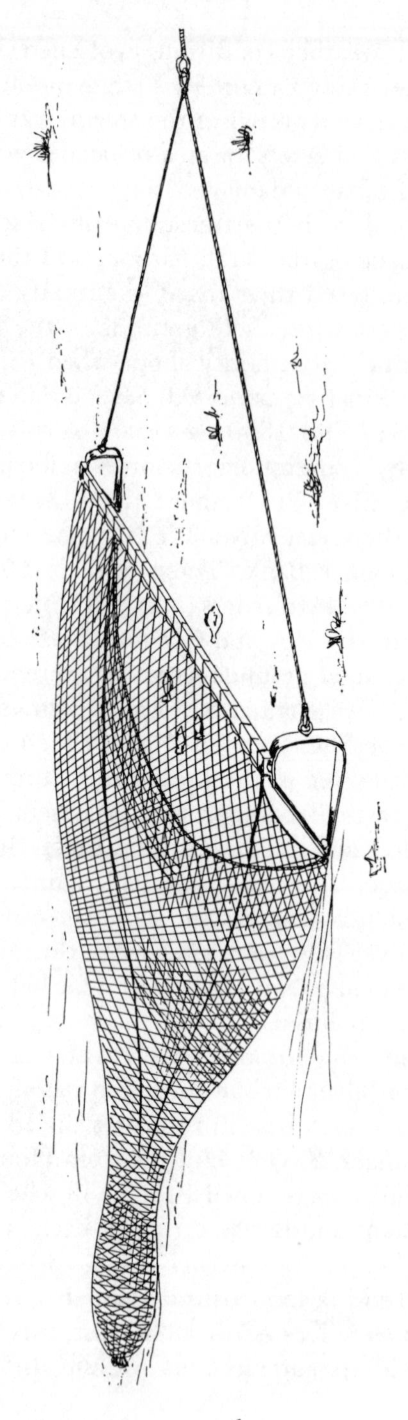

Fig. 2. The Beam Trawl

forward. Yet it seems unlikely that experienced seafarers, having once located and worked the grounds, would be unable to find them in future years. More probably, their whereabouts were kept a secret by the first handful of smackmen to settle in the area—such practices are by no means unknown today. Alternatively, if it was William Sudds' smack that had discovered them in 1838 then he may have taken his secret back to Margate and not made further use of it until 1844. By that date Hull had its own rail link and Captain Laws' railway rating arrangement would have made it easier to capitalize on the supply.

Recent research has shown that large-scale exploitation of the Silver Pits began in the winter of 1844/5.[5] These grounds always yield their best catches in severe winters—known as Pit Seasons—when soles congregate in deep waters. Their value had evidently become common knowledge towards the end of 1844 during such a spell of intensely cold weather. A number of smacks from the south had stayed over after their usual summer season to take advantage of the high prices prevailing at Hull and Scarborough. Once the Pits—known at first as the Silver Banks—began to yield their rich harvest of soles, cities such as Leeds and Manchester felt the benefit. Normally, these fish were a luxury item inland but they were soon retailing in Leeds at between 4d and 6d a pound and proving of great benefit to the poor at a time of dearth.[6] Their discovery was even mentioned in *The Times*,[7] and in January 1845 almost double the usual number of smacks were out working on the grounds off the Dogger Bank. A peak of some 18,000 pairs of soles were despatched from Hull on one day alone. In an effort to take advantage of the London market a number of smacks packed their fish in hampers and hailed southward-bound steamers who took their fish up the Thames.[8]

The 'discovery' of the Silver Pits occurred at a time when smack activity from Hull was beginning to increase. More fishermen were tempted to settle at the port and other crews extended their seasonal visits. However, this alone cannot have been the factor which brought about the subsequent rapid growth of the trawling industry, even in this area of north-eastern England. Sole booms in the Silver Pits were limited to cold spells and could not have sustained the smacks during the remainder of the year. Though an attraction, they were largely a bonus and not the major cause of trawling's expansion which was closely tied up with the railways.

There were a number of reasons why trawlers were eventually to benefit more from the railways than any other method of catching white fish. Trawling had always been a means of taking large quantities of fish but much of a typical North Sea smack's catch consisted of offal fish—including haddock and plaice—which, as previously noted, could not stand the additional cost of inland transport. The early trawlers visiting the Yorkshire and adjacent coasts had really only been seeking prime fish—hence their interest in the Silver Pits. Though some of their catch of offal fish was sold to linemen for bait, only a little was landed for sale with as much as four-fifths of the catch being heaved overboard.[9] This practice earned trawlermen the wrath of the linemen who could usually dispose of most fish they hooked, and it strengthened their belief that trawling was harmful and destructive of stocks.

The Laws arrangement, of course, gave access to a somewhat wider market and made it worthwhile for trawlers working off the Yorkshire coast to land more of their offal fish. During the 1850s the marketing opportunities for trawlers at many English and Welsh fishing ports were greatly extended when the national rail carriage arrangements, conducive to the carriage of offal fish, came into operation. Henceforward, trawlers with their large catches of cheap fish were better placed to supply the new mass markets then, say, the great-line fishermen who had catered more for the quality market.

During the 1840s and early 1850s, trawling from the north-east coast continued to be based on the Yorkshire ports of Hull and Scarborough. At first Scarborough remained largely a seasonal base for the visiting smacks, but after it was linked to the rail network in 1846 a number of the port's yawls began to fit out for winter trawling on their return from the Yarmouth herring fishery each November.[10] Though two smacks had been previously registered there, both had since moved on to Hull but from 1850 there was a new spate of migration from the south to the port. First to arrive was William Toby. In May 1850 he registered the *Eliza*, a 47-foot one-masted smack, built at Plymouth and previously registered at Yarmouth. In July the *Providence* followed. She had been built at Brixham but last registered at Hull. Over the next year or so a number of others arrived, including one very elderly smack, the *Rover*. She had been built at Cowes back in 1793 and her owner was William Alward, a surname later to become synonymous with Grimsby's fishing industry.

The following month the James Westcotts, father and son, registered their smack, *Gypsy Queen*, which had formerly been based at Yarmouth. More southerners joined them over the following two years as Scarborough became established as a trawling station. An analysis of the Custom House shipping registers shows that these smacks had originated in distant boat yards. Of the twenty-seven registered there during the 1850s, only one was built in Yorkshire and that at Hull. Two others were from Yarmouth boat yards while the rest had been turned out on the south coast with Brixham and Rye accounting for half a dozen apiece.[11] Scarborough harbour had more than enough capacity to cope with this fleet, for it had been under-used by fishing boats and its coasting trade had declined since the arrival of the railways. Moreover, its capacity had been considerably increased in 1844 when the outer harbour, which had been previously almost unusable, was dredged for the first time.

Most of the new arrivals were mortgaged, a method of purchasing craft then almost unknown on the Yorkshire coast. The substantial sum required to purchase a large fishing craft, was usually raised by a partnership. A typical three-masted lugger, costing over £600, was owned by four people—perhaps three of the crew and a landsman. Ownership of each vessel was therefore diffused and the owners were entitled to a share of the craft's earnings proportionate to their holding.[12] In contrast, the Brixham trade operated the mortgage system. A typical smack in the early 1850s might cost up to £600 from the builders yard. A thrifty trawlerman might raise the cash to purchase a new or second-hand one by saving hard and raising a mortgage from some wealthy individual. The vessel would be used as security until the principal was paid back, the trawlerman paying an annual rate of interest of about five per cent. This system of individual rather than collective acquisition was common in the Devon trawling trade where the low cost of entry and the local availability of credit, in combination with the system of earning by share of the catch enabled thrifty and hard-working individuals to raise the necessary capital. A typical smack there was owned and commanded by the skipper who held no financial interest in any other vessel.[13] In a great many cases he would have gone to sea from the age of fourteen for seven years as an apprentice and then saved hard after coming out of his time. With these savings he would have negotiated a mortgage from a local merchant then raised credit to fit out the smack from the local tradesmen. To date, the mortgage system had tended to encourage individual ownership of each vessel in the South West, unlike the collective ownership found in Yorkshire. It had also meant that no individual or company had built up a large fleet of craft as was the case at Barking where firms such as Morgan's and Hewett's owned and directed substantial fleets of craft.

The southern influence in the Scarborough fleet began to fade from 1855, thanks mainly to efforts being made by the railway company at Grimsby to lure fishermen to that port. Many vessels moved as the docking facilities being offered seemed superior to Scarborough's tidal harbour. There were other factors: George Alward recalled later that his family's decision to move was due partly to the problems of entering Scarborough harbour in winter but also to the difficulty that outsiders experienced in gaining acceptance amongst the local fishing community.[14] Although trawling had taken firm root at the port there remained a strong element of opposition amongst some line and drift men.

From 1855 locals began to feature strongly in the port's trawling trade. The three key figures were Abraham Appleyard, the local harbour master and shipowner, together with James Sellers and Henry Wyrill, the latter of whom was the trio's only working fisherman. Back in the 1830s and early 1840s he had worked on merchant vessels in the Baltic trade but by 1845 he was part-owner of a two-masted yawl. Sheer hard work coupled with an eye for opportunity paid off and by 1859 he had interests in at least four Scarborough smacks. By that time he had also become well-established as a fish salesman and probably rarely went to sea himself. Being a fish salesman at that time could be a very remunerative occupation for they made their money by taking a percentage of the ever-growing volume of fish sales passing through their hands as railways opened up the market. Some invested this wealth in the catching sector and none was more successful than James Sellers. He had been born in Malton in 1820 and until 1845 worked with his father in the fish traffic to the town from the coast.

When the railway to Scarborough opened in 1846 he moved to the port as a fish salesman, developing a range of commercial connections with inland towns. In 1852 he bought his first smack, and went on to own a substantial share of the Scarborough fleet.

Unlike many traditional fishing vessel owners on the Yorkshire coast, Sellers and Wyrill not only promoted trawling but they also made use of the mortgage system. Yet unlike Brixham, where the system had primarily encouraged single-boat ownership, these Scarborough men used it as a means of building up their interests in a fleet of craft. By the end of the 1860s Sellers was the largest fishing vessel owner on the Yorkshire coast. Sellers and Wyrill also bought their way into other craft by acquiring the traditional half and quarter shares and developed interests in the line and herring fisheries.[15]

By 1863 there were thirty-five Scarborough vessels which trawled for at least part of the year. Of these only fourteen were specialist smacks. The others were dual or even triple purpose craft. Their seasonal round included herring fishing in the summer and they alternated between trawling and lining during the rest of the year. Though still known as yawls they increasingly adopted the gaff rig which was more suitable for trawling than the faster but less manoeuvrable lugger rig and could be worked by a smaller crew.

Attempts were also made by local ship-owning and commercial interests to establish Whitby as a trawling station. In 1849 one smack, the *King William* from Exeter, was briefly registered there; the stay of a second stack was also short but ended more tragically when the *Friends Goodwill* was lost with all hands. There were a few further registrations during the 1850s, one of which was also lost at sea, but none stayed permanently.[16] Most fishermen at other inshore north-east coast fishing communities continued to revile the practice of trawling with the exception of Bridlington Quay, where several elements encouraged its establishment. Firstly, the bay there offered a considerable area of soft-bottomed sea bed ideal for inshore trawling. Secondly, the town had no strong fishing community at that time; those wishing to trawl did not have to contend with the hostility of their neighbours. Some of those who took up trawling came from surrounding towns and villages whilst others were locals previously engaged in servicing the fleets of passing colliers which had been thinning in number as railway competition took its toll on the coasting trade. Trawling offered an alternative means of earning a living. The Bridlington Quay trawlers were the traditional cobles rigged with a fore lug and jib and usually crewed by two men. They hauled a trawl with a beam of between twenty and twenty-four feet during a season lasting from about February to October.[17]

With no established fishing interest to contend with, trawling rooted itself most quickly at Hull and, once established, grew vigorously. In the early 1840s most smacks continued to be only seasonal visitors with just a handful registering at the port. It was only after Joseph Todd registered his new Hull-built smack, *Ranger*, in February 1844—soon to be skippered by Samuel Decent[18]—that the smacks became officially based there in any numbers. Within the next fourteen months a further ten smacks had been entered at the Custom House and the number increased gradually until by 1849 Hull could boast a fleet of twenty-nine smacks[19] which were supplemented, of course, by a much larger host of of seasonal visitors.

By this time the seasonal visit had become much more than just an occasional part of the annual round of activity and smacks stayed for an ever-increasing length of time. Brixham boat insurance clubs made specific provisions to cover these voyages and smack crews not only brought their families along and rented them accommodation for the duration, they also shipped their household furniture. From this it was only a small step to permanent settlement and more and more crews made this move as the 1850s wore on. Other smacks remained only visitors as late as the 1880s and by no means did all trips north go smoothly. In 1863 Clement Pine of Brixham visited Hull and Sunderland. His trip proved a disaster for he lost all his fishing gear and was obliged to sell his smack to cover the expenses he had incurred. His crew dispersed and, alone on the quayside at Sunderland, he collected together and sold all his worldly goods. With the money he purchased a nineteen-foot undecked boat and determined to make his own way home. His provisions were sparse, consisting of no more than a quarter stone of biscuit, two pounds of bacon, one ounce of coffee and one-and-a-half gallons of water. He set sail from the Wear at midday Thursday 9 July and passed Hartlepool some twenty-four hours later. A strong south-east wind carried away his sprit in the Boston Deeps but, unable to refit, Pine pressed on to Dover which he reached on the Monday. His rations were supplemented by catching ling and haddock and these were cooked on an apparatus consisting of old iron hoops set on the boat's ballast stones. At Dover, some fellow trawlermen offered him a tow but by now he was determined to see his voyage through and sailed on, reaching Ryde the following Friday. His tiny craft left the Isle of Wight on the Sunday afternoon and was becalmed off Portland all night. On reaching Teignmouth he grounded on the bar and had to wait for a floodtide to get off before finally sailing on into Brixham harbour, some thirteen days and 600 miles out of Sunderland.[20]

A similar strength of individual spirit was channelled by others into building up Hull's trawling trade. Many of those early settlers in the port were to provide its backbone for several decades to come. Such pioneers included Richard Vivian, Thomas Halfyard, John Guzzwell, John Sims and William Markcrow. Most came from either Brixham, Plymouth or Ramsgate and its neighbours. Guzzwell, for example, who in 1851 owned a fleet of three smacks employing eight men and seven boys, was a Devonian. Most were self-made and, as at Scarborough, a number of them found they could use the mortgage system to help them build up a fleet of smacks. John Sims, another native of the South West, probably Plymouth, had come to Hull in 1845 as an apprentice on a smack. He bought his first smack, *Jane*, in 1854 and gave up the sea to concentrate on management after acquiring a second, *Kingston*, in 1858. In 1878 he is recorded as owning four large smacks and by 1883 was president of the Hull and Grimsby Smack Insurance Society. His story of 'going in at the hawse hole and coming out at the taff rail'[21] was by no means unique nor was his rise singularly dramatic. Joseph Potter arrived from Brixham as an experienced smackman, having been apprenticed to the trade at the age of eleven. By 1863 he owned and managed two smacks, having given up going to sea himself, and his fleet had grown to eight large smacks by 1878.[22] Such a rise required an ability to save and handle money as well as a great deal of hard work and self-discipline. Many of those who came to be smackowners were abstemious, for drink was said to be the undoing of many fishermen after a hard trip with few creature comforts.

Others who arrived in the 1850s included Richard Hamling and

Robert Hellyer, surnames long to be associated with the Hull trade. Not all who built up substantial holdings were practical fishermen. Alfred Wheatley Ansell was born at Ramsgate in 1834 and, though his father had become a smackowner after having run away to sea on a Gravesend cod smack and later saw action in the Royal Navy at Copenhagen, he himself had been apprenticed at the age of fifteen to a Hull draper and silk mercer. After nine years in the business he married the niece of a large smackowner and entered the trade by acquiring his first smack. He soon diversified, having for a while an interest in groceries and a seamen's outfitters before concentrating on smack-owning and the wholesaling of fish. By 1878 he owned eleven trawlers and had extensive interests across Hull's fish trade.[23] Neither were all owners men: Jane Witty, for example, had an interest in several smacks during the 1850s and 1860s.

Until the 1850s facilities for the fish trade were poor. The Hull Dock Company took little interest at first above demanding dues of 6d per ton. The smackowners considered this to be too high and generally moored in the roadstead either discharging their catches by small boat or landing on a congested riverside quay by Nelson Street adjacent to the ferryboat pier. Eventually the Dock Company were prevailed upon to lower their dues progressively and by 1851 these were low enough to encourage all smacks to use Humber Dock. There the trade was provided with shed accommodation for a fish market on the south-western quay. This area became known as Billingsgate and, though some of those associated with merchant shipping continued to regard them as a nuisance, Hull's trawling trade continued to prosper. The fleet grew rapidly during the 1850s and, even though an increasing amount of fish bypassed the port, being sent direct to London from the fishing grounds by fast sailing cutters, pressure on facilities became acute. The tonnage of fish landed there rose from 1,571 tons in 1854 to 10,782 tons some ten years later. By 1867 a staggering 260 smacks were using the port on a regular basis, yet only five could be berthed at the same time whilst Grimsby could handle fifty. Hull smackowners pressed for improved facilities and after protracted negotiations the trade was allowed to move to the new 24½-acre Albert Dock.[24]

Further down the east coast, trawling was almost equally of interest. The ancient herring port of Yarmouth had been an occasional base for smacks since earlier in the century, but the

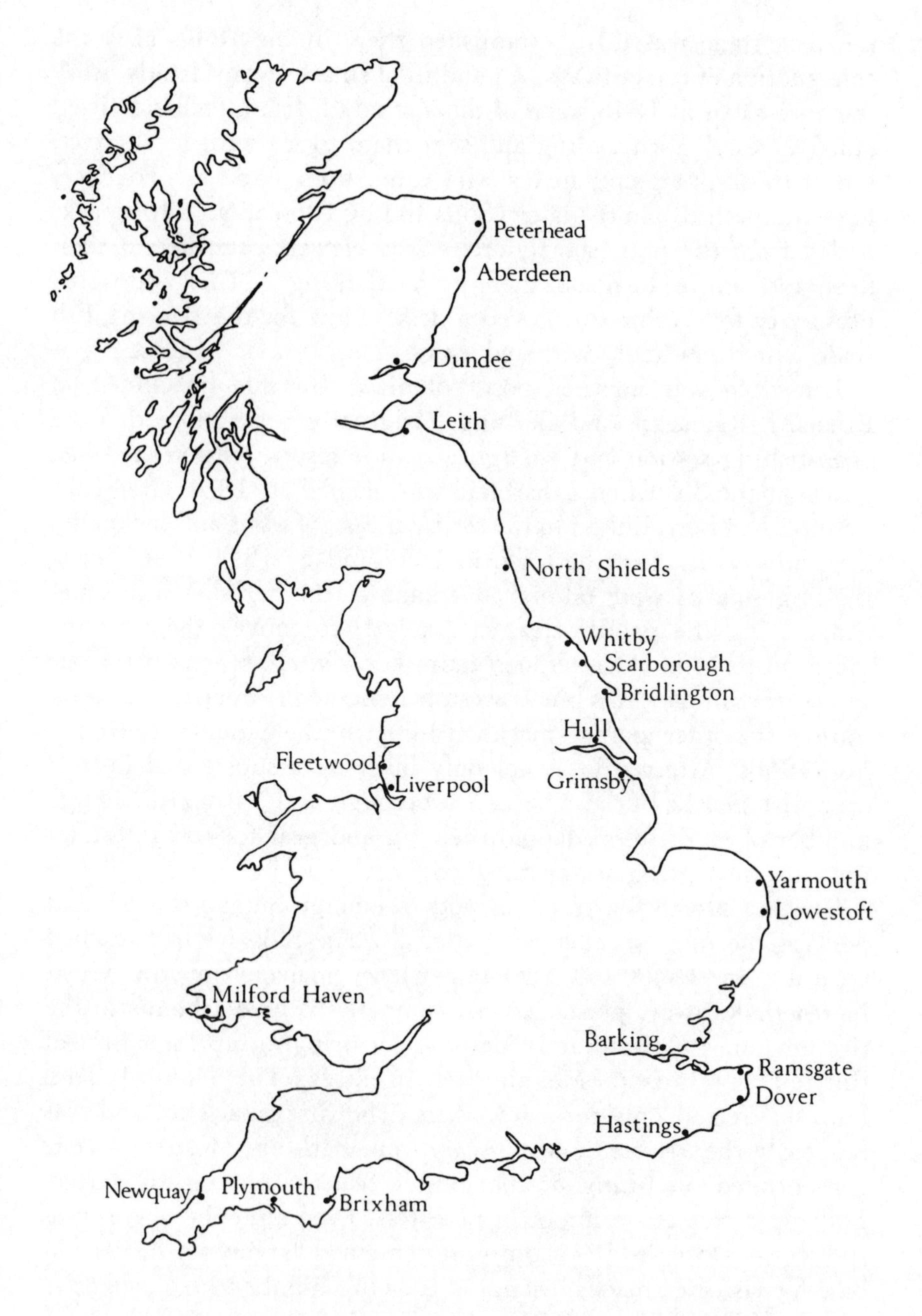

Map 1. Trawling Ports of the British Isles, 1900–1950

trawling trade was really established there in the 1840s after the construction of the railways. A handful of smacks were already using the port when in 1848 some of the Yarmouth fish merchants fitted out five vessels for trawling and sent them to sea with local crews, but with skippers and mates who came from Barking. The fleet grew dramatically in the later 1850s and by 1864 about 140 smacks hailed from the port; shortly afterwards Hewett's transferred their fleet and entire business there from Barking.[25] This additional prosperity for Yarmouth proved a death blow for the Barking fish trade which gradually withered away.

Lowestoft was another port colonized by the smackmen of Brixham, Ramsgate and Barking. Originally little more than an open-fishing station and emerging seaside resort, its potential had grown markedly when a harbour was opened in 1831 after Lake Lothing had been linked to the sea by means of a cut and lockgates. The railways reached the town in 1847 and by 1851 a number of Barking smacks were taking advantage of the harbour and sometimes using the paddle steamer *Lowestoft* to convey their catches ashore in time for the afternoon train. Local interests soon fitted out several trawling vessels but Lowestoft remained primarily a landing station for other port's smacks fishing in the vicinity until the mid-1860s. Afterwards, it not only built up a substantial fleet of specialist smacks but also, as at Scarborough and Yarmouth, a large number of its drifters adapted their rig and gear for trawling after the autumn herring season finished.[26]

The last North Sea trawling port to emerge during this era had perhaps the most spectacular rise of all. The railways had reached Grimsby in 1848 and Seymour Clark, manager of the Great Northern Railway, persuaded his company to give a bonus to the firm of James Howard and Company for bringing up their fleet of thirteen vessels from Manningtree in Essex. The Howard fleet, however, stayed only for a few years. The first smack to land was reputedly the *Princess of Wurtemburg* from Barking. Most early craft concentrated on lining or the lobster fishery, and Grimsby soon built up a tidy trade importing lobsters. Even after the opening of the Royal Dock in 1852 the port remained largely an occasional base for visiting smacks, though a growing number were using the port by 1854. In terms of its situation Grimsby appeared to have certain advantages over Hull. Being at the mouth of the river, the live fish brought in by the well-smacks could be kept alive in the water at Grimsby whereas they died upriver at Hull. Smacks using Grimsby could also save a fifteen mile run up the Humber, and claims were made that craft could land their fish and be back on the grounds in the time it took to sail on to Hull.

The Manchester, Sheffield & Lincolnshire Railway recognized this potential and began the construction of a fish dock in the summer of 1855. At the same time they set about marketing their venture by inviting the fish trade to a meeting in the Yarborough Arms at New Holland in June 1855. In order to entice smacks from other centres they offered a range of incentives including favourable dock charges, a wide range of attractive railway charges and free journeys for fish merchants travelling to inland towns to establish commercial connections.[27] Grimsby's fish dock was completed in March 1857 and a number of smacks lured there from various ports moved in. Only about ten Hull smacks moved at that time but others came across in the last few months of that year. Their owners included John Gidley and George Jeffs, names soon to be well known in Grimsby's trade.

After the first fish dock opened the Grimsby trade went from strength to strength. In 1858 the railway company built twenty-five fishermen's houses and within a short time rapid urban development was underway. A thatched ice house was constructed followed by accommodation for a range of allied trades including fish processing, boat building, sailmaking, netmaking, blacksmiths and twine merchants.[28]

By the end of 1863 there were at least seventy-smacks and forty-two line boats based at Grimsby[29] and many craft from other ports were using its facilities on an occasional basis. The trade's vitality showed no signs of diminution and a second ice company was formed in 1864. Within a couple of years of the fish dock being opened, the smackowners and fish merchants were calling for its extension and in 1866 the MS&LR agreed to do this. Another lock pit was added to ease congestion in 1869. The town's economy was soon dominated by fishing and it has been estimated between a half and two-thirds of the inhabitants owed their livelihood to the trade.[30] In 1871 the number of fishing vessels registered equalled those at Hull, and Grimsby soon established itself as not only Britain's but also the world's premier fishing port. Yet this rise was not at the expense of its neighbour, for Hull continued to grow and closed the gap again in the steam-trawling age. Between them the two Humber ports were at the centre of the world's trawling trade as the 1870s opened and yet thirty years earlier the river had had no fishing fleet to speak of.

Though the focus of the trade had moved with the establishment of the new trawling stations, much of the capital, enterprise and labour that fuelled this initiative originated on the south coast. The 1851 Census shows, for example, that there were already more than 1,000 persons from Cornwall, Devon and Kent living in Hull.[31] Although the south-western trade, for example, was increasingly overshadowed by developments in the North Sea it nevertheless still grew vigorously. Moreover the district also continued to contribute to the more modest expansion of trawl fishing from the Irish Sea ports of Dublin, Fleetwood and Liverpool. The national influence of Devon and, indeed, the entire south-coast trawling trade was probably at its height during the 1850s. Not only did many of those who opened up Hull, Grimsby and other places learn their trade at places like Brixham and Ramsgate, a great deal of the capital embodied in the catching effort came from the same districts. An analysis of the 168 smacks registered at Hull during the 1850s reveals that forty-two had been built at Devon boatyards. Brixham led the way, with Upham's yard playing an important role, whilst other smacks came from builders at Plymouth, Galmpton and Kingswear. A further eighty-four craft had been built in the south-eastern ports which Devon men had opened up in earlier decades. Hoad Brothers at Rye were to remain prolific suppliers of smacks to Hull for several decades whilst many other vessels came from yards in Ramsgate and Sandwich. A great deal of the capital which provided mortgages for the Hull pioneers also originated on the south coast.[32] Though the influence of the Devon yards was to reduce after the 1850s, as Hull and district boat builders came more into their own, Rye and Sandwich were to remain major suppliers of North Sea smacks.

5

Free Trade and Indentured Labour

By the mid-1860s English trawlermen could look back on a twenty-year period which had transformed their trade from a regional to national activity. The last ten of these had proved particularly prosperous, being marked by an unprecedented rate of growth which showed little prospect of diminishing. Fleets of trawlers were sailing into the North Sea from the Thames and Humber as well as Yarmouth, Lowestoft and Scarborough. Some 140 smacks worked out of the Devon ports of Plymouth and Brixham whilst Ramsgate boasted fifty more. Another 120 or so could be found in the Irish Sea trawling from ports such as Dublin, Fleetwood and Liverpool. There were in total nearly 1,000 smacks, manned by over 5,000 souls, and representing a capital investment of over £1 million. This fleet supplied the market with over 350 tons of fish daily.[1] The trawling trade, in short, was booming.

The diffusion of this controversial practice inevitably affected the activities of more traditional fishing communities, who viewed trawling with disdain and often reacted with outright hostility. Relations between the two groups worsened during the 1860s, especially as the trawlermen encroached on more and more grounds worked for countless generations by traditional fishing communities along the north-east coast of England. Fishermen from these places made frequent claims that trawlers over-fished the stocks, landed poor quality fish and damaged the gear of the line and driftermen. The lack of available statistical information or of any national organization specifically responsible for monitoring the progress of the fisheries made such assertions very difficult to prove or refute.

Prior to 1850 officers of the Edinburgh-based Board of British Fisheries had reported and produced statistics on many districts of the British coast.[2] The Board's prime function was not to superintend the fisheries but to supervise curing standards and this led it into conflict with the growing free-trade lobby which rejected such state interference in the workings of the economy. In 1848 John Shaw Lefevre conducted an inquiry for the Treasury which concentrated on its role in the curing trade. Lefevre found 'the system of authenticating the quality of goods by agency of a government officer . . . objectionable in principle' and the Treasury, acting on his recommendations, reduced the scope of the Board's activities, confining it mainly to Scotland.[3] Supervision of the fisheries was left largely to a few naval craft on coastal patrol duties. Without effective superintendence many fishery laws were never properly enforced. Fishing was supposed to be banned on a Sunday but many vessels continued to work over the weekend. A trawl's beam was limited by law to thirty-eight feet in length but by 1863 many smacks carried beams of forty-two or more feet.[4] Minimum mesh sizes were also rarely enforced.

Though many smacks fished well out to sea, others spent time from the late 1850s exploring the potential of more northerly grounds closer to the coast and were soon regularly working up as far as Newbiggin in Northumberland. Attempts were also made to trawl in the approaches to the Firth of Forth. These operations only increased anxiety and resentment amongst the traditional fishing communities who sought protectionist measures from the Government. Towards the end of 1862 a concerted attack on trawling began. Meetings were held in many communities from Newbiggin to Hartlepool, followed by others at Runswick, Whitby, Scarborough and other places along the Yorkshire coast.[5] Trawling's opponents were divided into those who were determined to see it totally banned and those who wished to see its operations restricted, usually to over twelve miles from the coast. During the following months MPs were lobbied and Parliament petitioned. This campaign succeeded in provoking a reaction: further meetings in support of trawling were organized in Brixham, Hull and similar ports, and the trade lobbied local MPs.[6] The general call was for an inquiry which both sides felt would vindicate their position. The storm spurred Palmerston's Government into action and in June

1863 a Royal Commission was instituted to ascertain the following matters:

1. whether the supply of fish was increasing, stationary or diminishing;
2. whether any of the methods of catching fish involved wasteful destruction of fish or spawn; and if so, whether legislative restriction would increase the supply of fish;
3. whether any existing legislative restrictions operated injuriously upon any of the fisheries.

The three commissioners appointed reflected to a fair degree the dominant scientific and economic orthodoxy of the time. George Shaw Lefevre's father John had already been responsible for confining the Board of British Fisheries to Scotland. James Caird, writer and agriculturalist, had played a prominent part in the free-trade controversies of the 1840s and had been an ardent supporter of the Manchester School. Thomas Henry Huxley, the chairman, was one of the most eminent scientists of his day. This brilliant and largely self-taught man was noted for his sound defence of Darwin's theories on evolution as well as his own work in the field of biology.

The Royal Commissioners visited most parts of the British Isles and interviewed a wide range of people connected with the industry. Their search for answers was hampered in part by a lack of suitable data which meant relying excessively on evidence gleaned verbally from witnesses as they travelled the coasts. Not unnaturally, much of this was subjective in nature and offered widely differing views, being often in line with the interests of the witnesses. For example, the bulk of the evidence given by the Yorkshire and north-east linemen suggested that the size of landings was on the decline whilst the trawling trade at Hull and Grimsby claimed they were on the increase. The Commission's other problem was that the science of marine biology was still in its infancy. There was little knowledge about the activities of fish apart from the fact that they produced millions of eggs. Not surprisingly, much of the biological evidence it accepted was incorrect. Huxley certainly believed that fishes were so prolific and the sea in which they swam so large that the activities of fishermen could have no real effect on stocks.[7]

That the fishing industry had undergone a major expansion was hardly in doubt. True, many traditional fishermen now faced competition on grounds worked by their ancestors for generations, but unlike the farmer's field the sea bed had never been appropriated and could be freely fished by anyone. In the Victorian age of *laissez faire*, attempts to restrict such freedom were not likely to be looked on with favour. There was no doubt that trawling was a formidably efficient means of taking white fish, and that efficiency, underpinned by a biological theory which held that the fisheries were inexhaustible, won the day.

The Commission was unfortunate enough to sit before the sheer scale of operations and the arrival of the steam trawler made even the trawler owners concerned about the future of North Sea fish stocks. They concluded that 'the supply of fish obtained upon the coasts of the United Kingdom of late years has not diminished but has increased'. But one crucial point which the Commissioners were unable to examine effectively in the absence of much hard data was whether the continued increase, in terms of boats and gear deployed, was accompanied by a commensurate increase in the amount of fish actually caught. In other words, whether marginal returns in terms of fish landed for each unit of capital and labour deployed were constant, increasing or decreasing. The situation they faced was confusing, for smackowners claimed that trawlers were landing more fish. One later stated his belief that the action of the trawl cultivated the sea as the harrow does the land whilst a Billingsgate fish merchant told the Royal Commission that the trawl operated on the soil of the sea as did the plough on the land.[8] Yet by 1863 smacks were landing most of the offal fish they had previously thrown back and this made comparison with earlier years most difficult. Moreover, many of the new smacks were much larger—and deployed larger beam trawls which increased their efficiency—than were the craft which had opened up the North Sea some twenty years earlier. In contrast, the linemen on the Yorkshire and north-east coast claimed that the yield per fisherman and boat was decreasing but the Commissioners observed that the numbers of such fishermen was continuing to increase. However, it seems quite possible that the traditional line fishermen were buoyed up by the higher quayside prices as the railways revolutionized the nature and scope of the inland market.[9] It is conceivable that a smaller catch in 1863 was worth more than a much larger one landed some twenty years before.

The Commission reported in 1866 and its first recommendation was that 'all Acts of Parliament which profess to regulate, or restrict, the modes of fishing pursued in the open sea be repealed; and that unrestricted freedom of fishing be permitted hereafter'. It made similar recommendations for inshore fisheries but with certain conditions. Not all of the recommendations were directly related to trawling but covered, amongst other things, the policing of the fisheries and a call for the systematic collection of statistics to prevent the 'constant recurrence to panics to which the sea fishery interest has hitherto been subjected'. Finally, it recommended that all restrictions preventing foreign fishermen from entering British ports be removed and that a similar freedom be sought for British fishermen. So far as the fishing industry is concerned, this report has been described as 'the true and final apotheosis of classical *laissez faire*'[10] for it advocated the freedom for all to fish anywhere and by any means, and to sell their catches without restriction or assistance, irrespective of nationality. The Government followed many of these recommendations and trawling was free to make the most of a free market. It continued to grow at a phenomenal rate and by 1878 the two ports of Hull and Grimsby alone had nearly 1,000 smacks on register.[11]

There remained one factor which restricted the rate at which the trawling trade could expand and that was the availability of labour. Traditionally, fishermen looked to recruit young crew members from amongst their family and close neighbours. Entry was from within, and the youngsters became apprenticed to their fathers, uncles or close friends, as still happens in many small ports around the British Isles today. This system resulted in the development of many stable and close-knit fishing communities. Recruitment from within existing communities, however, was insufficient to sustain the rapid expansion of trawling, particularly at the new ports of Hull and Grimsby.

Many apprentices had long been recruited from amongst the young poor and underprivileged. The Elizabethan Poor Law Act of 1601 directed that destitute children should be apprenticed to a trade; such a system had laudable motives and in many trades was used as a means of allowing youngsters to acquire the necessary skills with which to earn their livelihood. Yet in some occupations the system was open to abuse, providing less scrupulous employers with a means of acquiring cheap labour. By the nineteenth century some parish officials saw apprenticeships more as a means of easing the burden on ratepayers than as a way for a youngster to acquire an appropriate skill, as readers of *Oliver Twist* will be all too aware.

The terms of the indentures usually bound an apprentice trawlerman to his master until the age of twenty-one. For his part, the master engaged to provide his apprentice with clothes, food and lodgings as well as medical assistance. Though the master was not obliged to make any substantial monetary payment, the lad was generally kept in pocket money or allowed to make something on the smack's catch of stocker bait, which consisted of gurnards, rays, monks and dabs. When an apprentice satisfactorily completed his term of servitude it was customary for the master to present him with three good suits of clothing.[12]

Traditionally, the apprenticeship system had ensured that the trawling trade secured a supply of experienced and competent fishermen. Though an apprentice's lot in any trade depended acutely on the character of his master, as Crabbe's melancholy tale of Peter Grimes illustrates all too clearly, many early trawlermen were probably contented with the way they had entered their calling. The master had generally lodged them in his own house, supervised their conduct, and otherwise acted towards them in *loco parentis*—this became known as the indoor system. As we have seen, many former apprentices were to rise to the very top of their trade in the last half of the nineteenth century.

As trawling spread and developed at ports such as Grimsby and Hull which had no tradition of fishing, the apprenticeship system began to change for the worse. The sheer numbers of smacks entering service from the 1850s through to the 1880s placed skilled labour at a premium. Most experienced smackmen could obtain either a skipper's or mate's berth and many of the fishermen at traditional stations like Staithes and Runswick could not be lured to the despised trawlers. Soon it was commonplace for the other three crew members to be apprentices and labour was sometimes so scarce that smacks occasionally went to sea manned only by apprentices. By 1872 the apprentices at Grimsby outnumbered the fishermen by 1,350 to 1,150.[13] The Merchant Shipping Act of 1854 stipulated no minimum age for apprentices and there were cases of lads being bound at the age of eleven for ten years servitude on the smacks. Nor was it considered necessary for any parent, guardian or other responsible person to see that youngsters fully comprehended what they were binding themselves to or that the master properly discharged his obligations. Pitiful reports occasionally appeared in the newspapers about parents who could do little when they found that their runaway children had been indentured to the smacks.[14]

In order that the smacks could continue to open up deeper and more distant waters their owners had to spread the recruiting nets ever further into the lowliest corners of English society. Many lads were indentured from poor-law union workhouses across the country. London proved a prolific ground: the Holborn Union supplied Hull with many boys over the years, as did the Stepney Union to Grimsby. Leeds was also a fruitful source and rural workhouses supplied many Yarmouth apprentices.[15] A smack-owner taking on a workhouse apprentice might receive £5 to help with his lodging and clothes from the Board of Guardians. Others came from reformatories, charitable institutions and various public bodies. Some were wanderers, waifs and strays, young inhabitants of the Victorian urban underworld.

Many smackowners relied on such recruits throughout the rest of the apprenticeship era: in 1894 a analysis of the background of the 1,420 apprentices indentured at Grimsby over the previous five years revealed that 715 had been apprenticed by workhouse guardians, seventy-four came from reformatories and industrial schools while another sixty-eight were recruited from Greenwich Hospital schools, training ships and missions.[16] All in all, some 857 had been apprenticed by public bodies. Many others had run away from home and the sea had always been a place to run to. Yet even before the 1860s were out the labour shortage had become so acute that the smackowners resorted to an almost wholesale system of recruitment with suitability and aptitude apparently being of low priority.

A large proportion of lads recruited from these sources were unruly, unhealthy and in need of the sort of close supervision that the indoor system could provide. By the late 1860s, however, this method of providing for apprentices was on the decline. As smackowners built up fleets of craft and became more affluent, they became increasingly reluctant to keep this growing army, drawn from the streets and slums, under their own roofs. Some took the trouble to find them suitable lodgings but others adopted the so-called outdoor system whereby the master dispensed with his obligation to provide lodgings and food for the lads on shore but gave them a financial allowance instead. In other words they were left to fend for themselves, and this trend merely exacerbated the situation. The lad's main duty to his master was to be back on board at the time set for the next voyage. On landing after the cold, hard, grey toil of a trawling trip, apprentices of all ages immediately sought the contrast of crowds, colour and warmth to be found amongst the bright lights and laughter of nearby public houses and music halls. Little about their life encouraged a long-term view. With sometimes only a couple of days between trips, many youngsters made the most of their freedom and spent their allowances on an immediate orgy of excess. Grimsby apprentices were often regular drinkers from the age of fourteen years and commonly roamed the streets or stayed in low boarding houses frequented by all manner of characters. It was reported from Grimsby in 1882 that lads as young as sixteen kept girls in these garrets and lived together as man and wife, referring to each other as 'pals' or 'chums' respectively.[17]

The situation at Hull was little better, and as the money they could make from their share of the stocker-bait perk grew during the 1870s—especially as trips became longer—they increasingly fell prey to low life on the waterfront. Both indoor and outdoor apprentices were tempted but the latter always had the most money and freedom. In 1882, the approaches to Albert Dock were reported to be 'infested' with young 'wenches' intent on earning their living from the lads who, coming up from a long trip with as much as £3 in earnings, were sometimes all too eager to take advantage. Lads were reported to be spending their time ashore in a semi-drunken state in brothels in the vicinity of Trundle Street and Union Court off Waterhouse Lane.[18] Not surprisingly, venereal disease was rife amongst apprentices in many ports and some smackowners complained of the expense this caused them in medical fees.

The freedom of these brief forays ashore contrasted starkly with the rest of an apprentice's seven or so years of servitude. Most of his time was spent afloat in conditions that had always been spartan and extreme. Life on the smacks became more arduous as voyages lengthened and operations intensified. Trawling was always dangerous and no less than 304 men and boys lost their lives on Grimsby smacks between 1876 and 1882. Apprentices were always in greater danger because of their relative inexperience.[19] Large numbers of the lads being sucked into the trade came from the country's interior and had no previous experience of the sea. Many had probably never even seen it before their arrival on the quayside. Often a trial trip was undertaken before indentures were signed but many a first voyage through winter storms on a wet and constantly heaving smack can have been little short of terrifying and at least one lad died of fright.[20] For some others it was so awesome that they found themselves almost literally torn between the devil and the deep blue sea and preferred to be sent back to the hated workhouse and a life of 'less eligibility'.

An apprentice usually commenced service as the smack's cook and, as he gained experience, graduated to being a deckhand. Though firm discipline was an ingrained tradition of seagoing life, many lads were undoubtedly treated fairly throughout their years of servitude by the standards of this harsh trade. Most skippers were basically considerate men but some, conditioned by hard years before the mast and keen to make enough to acquire their own vessels, had little patience with the increasing number of apprentices who created problems. Moreover, the confined conditions under which the crew of five lived and worked for up to eight weeks at a time tended to shorten tempers, and even minor incidents could provoke strong reactions. In a number of notorious instances, the treatment of apprentices degenerated to the level of sadistic or sexual brutality.

It is hard to gauge the level of such cruelty on voyages far out to sea and yet a disturbing number of incidents were reported in the press from the 1860s onwards. Many of the lads recruited had little sense of personal hygiene and their appointment as cooks could be a source of intense aggravation. One notorious case of manslaughter concerned Jacob Kiesler, an apprentice on the smack *Comet*, whose skipper was Thomas Hamling. On his first trip—out of Hull at the end of 1864—Kiesler was berated by the crew for his particularly filthy habits. Soon afflicted by salt-water boils, he refused to work for a week, and incurred the particular wrath of the mate John Anderson who beat him on several occasions. At one point it was alleged that he was made to eat his own excrement in the presence of the whole crew, and on another that he was stripped and scrubbed with a whalebone brush. The lad later took a turn for the worse and died. On arrival back in port the police surgeon examined the body and at the subsequent inquest gave evidence that it was covered in bruising and swelling with a large contusion to the head. Death was apparently occasioned by extreme exhaustion, want of sufficient nourishment and ill-usage. The inquest jury returned a verdict of manslaughter against John Anderson and censured the conduct of the skipper.[21]

There was an increasing number of incidents in the 1870s, and several Grimsby skippers were fined for assault betwen 1875 and 1878. There were also a disquieting number of suicides by apprentices at sea. In August 1873, for example, the Grimsby smack *Gleaner* was working on grounds some thirty miles from Flamborough Head when Frederick Donker, the apprentice cook, was lost. He had been called down into the cabin where most of the crew were having their breakfast, and shortly afterwards was alleged by the apprentice on the tiller to have rushed back on deck, wringing his hands in agony and screaming, before throwing himself over the side. Grimsby was soon rife with rumour concerning his death and in September the master William Brusey, aged twenty-four, admitted beating the lad, was charged with manslaughter, and committed to Lincoln Assizes. At his trial, however, the judge expressed doubts over whether the prisoner had a case to answer and the jury dismissed the bill against him.[22]

Many apprentices were treated by the smackowners as their personal property. They were bound to service and if necessary could be used as a means of breaking strikes. During a trade dispute at Hull in April 1852 a smackowner, Mr Nichols, was summonsed by his apprentice for not feeding him. Moreover, the apprentice declared that he was refusing to sail on his master's smack whilst its crew was composed entirely of apprentices as he felt his life to be in danger. The magistrate, who was also the Mayor, sympathized with the lad but said that the bench could not intervene as his indentures bound him to obey his master's commands.[23] In other instances lads could be hired out without their consent. One woman who complained to the Board of Trade that she did not know her son had been bound was told that the lad, aged fifteen, could bind himself. His Hull master later lost his smack and then hired out the lad to another man for eighteen shillings per week. Six months later the master acquired another smack and brought the lad back but within a few months he was washed overboard and drowned.[24] Many such apprentices were bound into servitude within a mile of the statue of William Wilberforce, the Hull-born emancipator of slaves. As the fleets grew in size, the smacks shipped an increasing proportion of apprentices, but there was no guarantee of greater reward for extra responsibility whilst under indentures. At Brixham it was pointed out that an apprentice might get nothing for acting as a skipper.[25]

The apprenticeship system came to rely heavily on the courts and prison. Desertions were commonplace and under the terms of the Merchant Shipping Act of 1854 owners did not require a warrant to arrest all who deserted their smacks. Yet for some, gaol was little deterrent. Henry Webster, Governor of Hull Prison, told an 1882 inquiry that many would rather be in prison than on the smacks. One apprentice, Charles Jordan, agreed that a ship was a prison with an additional chance of being drowned.[26] Public disquiet about the increasing instances of imprisonment had manifested itself back in 1873 when the Lincoln Chronicle observed that:

> Lincoln rings with indignation at the treatment these lads receive at the hands of the authorities. The lads are brought by train, which generally arrives about 9.30 pm and are heavily chained together, in numbers from three to five, and in this way are marched through the busiest part of the High Street of our city for more than a mile to their destination.[27]

There were broadly two forms of desertion. Firstly, there was the offence of 'stopping the ship'. This meant that the lad did not turn up when the smack was due to sail or that he ran ashore as it was making ready to leave port. This was a costly practice for it usually caused a vessel to miss the tide and left the crew to lie idle whilst the offending lad was brought back or a replacement found. 'Stopping the ship' was a problem at all trawling ports and was considered to be particularly bad on a Monday morning at Brixham. Many apprentices used this as a means of prolonging their foray ashore whilst others used it to draw attention to grievances. A number deserted frequently, in the hope that their master might eventually lose patience and cancel their indentures.

The other, more serious, offence was that of absconding. An absconding apprentice intended to run away from his master. Many older apprentices ran off to other ports where they could earn wages as weekly hands. Others were so sick of their livelihood that they attempted to escape from the trade altogether. These lads had great difficulty in effecting a successful desertion: without money for fares it was hard to get away from ports like Hull or Grimsby by train, and so many lads attempted to run away that the authorities were always on the lookout for suspicious youngsters. Often the only way was to tramp the roads or ship out on a foreign merchant vessel. Others were imprisoned for bad behaviour at sea with several being brought before the magistrate for cutting the trawl warps which forced smacks back to port. Others mutilated themselves in the hope of being sent back on the next cutter.

By the late 1870s many of the magistrates they came before in the large trawling ports had interests in the fish trade themselves and the justice they doled out was often far from lenient; three months hard labour being a common sentence if an apprentice refused the option of returning to sea. In 1876 Hull magistrates sent some 216 apprentices to Hull gaol and their Grimsby counterparts despatched 221 to Lincoln the same year.[28] An analysis of Grimsby apprenticeship records after 1880 reveals that over twenty per cent had at least one term in prison and the figure was probably somewhat higher in the previous decade. Not all trawling ports resorted so readily to the law when dealing with their apprentices. Though some apprentices were imprisoned at Brixham for similar offences many more disputes were settled domestically.

Rumours about the treatment received by some apprentices had led to a number of formal and informal inquiries at Grimsby during the 1870s. In 1873 the Bethnell Green Guardians sent a committee to the port which noted that 1,800 boys had already been apprenticed to the trade there over the years. Although it found some complaints to be justified and made criticisms of the lack of supervision for many lads on shore, its report by and large approved of the system. Yet Grimsby in particular continued to have its image tarnished by tales of death and cruelty and reports of a growing number of apprentices spending time in prison. For a while the supply of recruits from the poor-law unions dried up and in 1878 another inquiry was made into the port's system, this time by Mr

Stoneham and Mr Swanston on behalf of the Board of Trade. Much of the evidence presented to them by the fish trade came from the smackowners and many of the apprentices who spoke found favour with the system. The report was generally favourable and found no evidence of systematic cruelty but again criticized the want of proper supervision for apprentices on shore. Acting on its recommendation the port's Mercantile Marine Office was separated from the Customs, and an additional office was established on the docks where the superintendent could ensure that the apprenticeship laws were being adhered to.

But by no means were all of the smackowners in favour of the system. Before the 1878 inquiry James Alward of Grimsby had spoken of imprisonment as a great evil and maintained that an inquiry would only expose their 'own rottenness'.[29] Indeed, outside of the trawling ports there were far fewer youngsters being despatched to prison. Some attempts were made to improve the supervision of apprentices ashore and these met with varying degrees of success. At Ramsgate many masters paid fourteen shillings a week to lodge their lads in the Fisherlad's Institute [30] and the official inquiries found less of a problem both there and at Brixham where there was a similar institution. The Grimsby Fisherlad's Institute was finally opened in 1880 but it only catered for a small proportion of the port's apprentices.

The whole system was thrown into chaos by one section of the Merchant Seamen (Payment of Wages and Rating) Act of 1880. Though this allowed owners to convey deserting fishermen aboard their smacks they were no longer allowed to arrest them without warrant or hold them in custody whilst their cases were being heard. Imprisonment for desertion could not be inflicted and the apprentice was allowed to give forty-eight hours notice of his intention to absent himself. The smackowners were left with recourse to civil law but this was of scant use when apprentices had so little property. Magistrates at most ports interpreted this closely and as soon as changes in the law became general knowledge there were wholesale desertions by the apprentices. By 1882 the number of apprentices still adhering to their indentures had dropped by seventy-five per cent and the system rapidly declined at most east-coast ports. In 1878 Hull had possessed some 400 apprentices but the number had fallen to sixty-two by 1893.[31] At Grimsby, however, the magistrates interpreted the law somewhat differently and continued to imprison deserting lads, apparently on the tenuous grounds that they had disobeyed lawful orders.[32] Though the numbers indentured at the port each year began to fall, the system survived. But if apprentices did manage to escape it was found that they could rarely be brought back.

Just as the smackowners were trying to come to terms with this Act they were thrown into the national spotlight thanks to the horrendous murder of two apprentices. In May 1882 Osmond Otto Brand, skipper of the Hull smack *Rising Sun*, was sentenced to death at Leeds Assizes for the murder during the previous December of fourteen-year-old William Papper. The third hand, Frederick Ryecroft aged eighteen, was also found guilty of common assault and sentenced to three months' hard labour. Papper had incurred the wrath of the skipper quite early in the voyage and was subjected to a course of beatings and cruelty whilst being deprived of food. The boy subsequently lost consciousness and though Brand tried to bring him round by forcing hot tea and tobacco into his mouth and even hung him over the side, he was never revived. Brand had his body thrown over the side and told the other apprentice that the lad had fallen overboard. The death was not reported at the police station when the craft came back to Hull and only came to light when a conscience-stricken crewman revealed the sordid tale.[33]

In February of the same year, Edward Wheatfield, second hand of the smack *Gleaner*, murdered Peter Hughes. The skipper, Daniel George, seemed to have little control over Wheatfield who subjected the apprentice to systematic beatings with both rope end and boot. Hughes was stripped naked and made to walk the decks with a bucket of water on his head. He was denied food and on one occasion was persistently booted by Wheatfield for three-quarters of an hour by which time the lad's hands were bare to the bone. Another crew member daubed the lad in his own excrement and later Hughes was thrown over the side. This case also did not come immediately to light as Wheatfield at first claimed the lad had fallen over the side whilst drawing water on board. Wheatfield was tried, convicted and subsequently hanged.[34]

These murder cases could hardly have come at a worse time for the smackowners who were seeking Government help to restore the apprenticeship system. The spotlight was thus thrown on their trade and several other alleged cases of cruelty brought before the courts received widespread coverage. Public opinion was outraged by the conditions under which the youngsters worked. In July 1882 a deputation of smackowners from the main trawling ports which went to see Joseph Chamberlain, President of the Board of Trade, received short shrift:

> Then you mean to tell me that one fourth of the people engaged in your fishing business break away from their engagements with their employers. Surely such a state of things does not exist in any trade or business. What can be the reason of it? Either your men do not like your bargains or they must be the very worst class of men to be found. What you say is this, that unless you have the power of summarily taking a man up and putting him into prison you cannot get him to carry out his bargain—that you cannot get men to work except under threats of imprisonment. That would be reducing matters to a state of serfdom. As to the apprenticeship system, I am not sure it is not a system more honoured in the breach than in the observance. We hear of the most cruel treatment of apprentices being common among smackowners in Hull and Grimsby, and I have been told by some of the largest owners in the fishing trade that they are not sorry the system is broken up.[35]

A Select Committee was set up to investigate the situation but many of its members clearly had connections with the ports of Hull and Grimsby. Its report was generally in favour of the existing system —an unsurprising outcome since its chairman had previously claimed that one of its tasks would be to find the best way of protecting owners against the losses caused by desertions. Acting on the basis of the 1878 and 1882 reports the Government passed the Merchant Shipping (Fishing Boats) Act of 1883. Amongst other things this gave wider powers to the marine superintendents in their role as guardians of apprentices, and charged them with ensuring that the lads indentured were over thirteen and had the approval of their parents or guardians.

Although there were attempts to maintain a viable apprenticeship system at Grimsby into the 1890s—and indeed the port still had a handful of lads as late as the 1930s—it could no longer cope with the expansion of the trade. East-coast trawler owners came to rely increasingly on weekly-paid hands over whom they had less immediate control. At the south-coast ports, however, where sailing smacks still operated, the apprenticeship system continued to thrive for some time. The system had less acute problems in those ports and the number of apprentices on indentures at Brixham rose from 210 to 248 between 1889 and 1893 whilst those at Ramsgate increased from 135 to 156 over the same period.[36]

Any worthwhile system of training provides trainees with an opportunity to obtain the competences necessary to be classed as 'skilled' in a particular occupation. Trades which required a high level of skill usually developed strong apprenticeship systems which the labour force were keen to maintain as a means of regulating entry. Scarcity of skilled labour placed those with the requisite skill in a stronger position when bargaining with employers. Employers in such trades were sometimes keen to dilute the system—perhaps by shortening the period of indenture or bringing in semi-skilled labour—as this tended to reduce the scarcity value and improve their bargaining position. In occupations where the apprenticeship system was used to try to guarantee a supply of cheap and acquiescent labour, the opposite tended to be the case. Here the employer was usually the keenest to maintain the system and this was the case with the fishing industry.

It could hardly be argued that the lads were drawn into signing indentures, which bound them to the age of twenty-one, because this was the only way of becoming a competent fisherman at the time. Whilst it is true that many fine trawlermen had been through the system it is also fair to say that many ports and other methods of fishing never employed the system and yet produced equally competent fishermen. The emphasis for employers seems to have been less on using the system as a means of acquiring a high level of skill than on the length of servitude. No matter what age a lad signed up, be he eleven or seventeen, most owners exacted the maximum period of indenture to the age of twenty-one years. Practical experience is acknowledged as an essential part of training, but little provision was made at the time to enable lads to back this up with relevant study in seamanship or safety. The quality of training received on the smacks was steadily diluted as the proportion of apprentices in the crews increased; someone who is considered competent to skipper a smack could hardly be regarded as a trainee. Later, more appropriate technical education and training was developed for all ranks of trawlermen.

In reality, the apprenticeship system became steadily less economic and increasingly troublesome. Not only the trade but also the courts and prisons had to devote ever-growing resources to the care and control of the community of apprentices. This detracted from the trade's primary purpose of seeking fish. The public and state disapprobation eventually reduced the flow of lads into the trade and finally forced the owners to use a somewhat freer labour market. It remains ironic that a trade which benefited so much from *laissez faire* on the high seas should have relied so heavily and for so long on such a tied and ragged labour force.

6

Fishermen and Fleeting

The 1870s represented the zenith of the sailing trawler. By the last years of the decade the activity was carried out intensively across the North Sea in waters to a depth of thirty fathoms and more. The banks adjoining the Dogger, including the Silver Pits and Botney Gut to the southward, were the main winter trawling grounds. Come summer many smacks made longer voyages and could be found frequenting waters close to the Danish, German, Dutch and Belgian coasts from the Hansholmen Light in Jutland southwards to Ostend. The Hull and Grimsby fleets usually worked the reefs north of Heligoland and began at the Horn Reef and Amrun Ground off the Danish coast about May or June before moving northwards towards the Little Fisher Bank close to the entrance to the Skaggerack. They returned to the neighbourhood of the Dogger Bank for autumn and winter fishing.

The Yarmouth, Lowestoft and Ramsgate fleets did not normally fish northwards of latitude fifty-five degrees. They mainly worked on southerly Dogger grounds in winter and during the summer trawled on banks south and west of Heligoland, including the Borkum, along with other favourite banks off Texel and the remainder of the Dutch coast. Brixham vessels also continued their nomadic habits. Between January and March most worked within twenty-five miles of Start Point or Berry Head but in March they dispersed, moving to either Penzance, the French coast, Irish or North Seas. The latter remained particularly popular and many Torbay smacks would land at Yarmouth or Lowestoft until the end of July after which they returned to work on the south-western grounds.

Plate 1. Late Nineteenth Century. A Ramsgate Sailing Smack at Work

The previous thirty years had seen considerable improvement in smack design. The most obvious alterations were to the hull dimensions and rig. An analysis of Scarborough smack registrations between 1850–1854 shows that forty-five feet from stem to stern was the average length: by the period 1875–1879 the average length had risen to seventy-four feet[1] and this trend was repeated—with regional variations—at other trawling ports. Smacks had traditionally been rigged as single-masted cutters but as they grew in size the larger boom proved difficult to handle. From about 1867/8 boat builders began fitting out smacks as two-masted ketches and this rig was soon universally adopted for all new craft. Bigger craft could ship more gear and by the mid-1870s the largest trawls incorporated a forty-eight foot wide beam[2] whereas thirty-eight feet had been commonplace in the 1840s.[3] Rope warps were replaced from about 1873 with stronger ones made of steel, enabling trawls to work into deeper water; at the end of the decade steam capstans were beginning to make an appearance. These not only freed the crew from the backbreaking task of hauling in the trawl by hand but, by speeding up the process, allowed them to make more shots in a given time. A number of smacks were also built with iron rather than wooden hulls. The Scarborough lining lugger *Contrast* was perhaps the first iron-hulled fishing craft. She was built in 1862 by Samuelson of Hull for Josiah Hudson and, although originally intended for lining on the Rockall Banks,[4] she was soon re-rigged as a trawler. Another early iron smack was the *Tubal Cain* which worked out of Grimsby from about 1870, but although an increasing number of such craft were built for owners at Grimsby, Hull and Scarborough during the following decade most continued to be constructed of wood.

A number of the older smacks were also lengthened so that they could carry larger beam trawls. At Scarborough they were altered on the open harbour beach. Everything that could be removed, including masts, spars and ballast, was stripped out and the smack was then 'neaped'—hauled out of the water as far as possible on a spring tide—before being cut in half at its greatest beam and pulled apart with the aid of block and tackle. A new section of keel was then fitted between the sections and the area above built up before the boat was made ready for refloating.[5]

These developments were accompanied by an inevitable rise in costs. A newly constructed trawling smack, fully fitted out, normally cost in the region of £700 to £900 in the early 1860s. By the 1880s the largest wooden smacks cost around £1,500 and as gear rose in price it became more difficult for the working fishermen to acquire his own new vessel, though many began by buying one of the numerous second-hand craft then on the market.[6]

Though smacks grew in size, the hazards of trawling on the unpredictable North Sea did not decrease. In December 1867, for example, fifty men and boys were lost when ten Hull smacks went down in a terrific gale. The relatively young age of those drowned probably accounted for the fact that they left only seventeen widows and twenty orphans between them.[7] Losses of fishing vessels were commonplace, and there were countless other occasions when a vessel arrived back in port to report at least one crew member washed away.

Statistics for the early years are inaccurate, for not all deaths were reported, but when they do become available they paint a stark picture. No less than 235 Grimsby fishermen were lost between 1879 and 1882 out of a total seagoing labour force of about 4,000. Between 1878 and 1882 some 109 were also lost from Hull.[8] The death rate was considered to be ten times that of mining. Not all losses of smacks could be attributed to storms. Many vessels suffered the fate of the Scarborough iron smack *Ocean* which was lost with two of the crew when run down by a steamer on the fishing grounds in March 1877.[9] The rest of the crew were picked up by the steamer and landed in Hamburg. Some hazards were perhaps less predictable. In September 1869 six Hull smacks working off the Dutch coast were lost after being caught in a great gale. Two collided with each other and the rest were eventually driven ashore on Heligoland. One of these craft, the *Priscilla*, owned by Henry Toozes, was considered the finest smack in the port and worth over £1,000. Three days after going ashore she was reported to be still sound and capable of being refloated. However, the crew alleged that as soon as they left and handed her over to the coastguard the islanders swarmed on board and stripped her of all moveable stores. When the news reached Hull the owner and insurance club despatched help but by now the *Priscilla* was so gutted that she raised only £26 for firewood.[10]

Fishing trips also changed as the smacks ranged ever further across the seas. As voyages had become longer ice was increasingly carried to keep the catches fresh. Ice seems to have first been used to keep herring and salmon on the voyage to London from Scotland at the end of the eighteenth century. Certainly, steamboats were conveying iced salmon to London from Edinburgh in 1838. Hewett's Barking fleet started the practice of taking ice to sea on smacks about 1847 and the innovation caught on. It became particularly important in the 1870s as longer voyages became more usual. Artificial ice was first used commercially by the fish trade in 1874 but it was not until the end of the nineteenth century that it became the chief source of supply. Before then the fish trade relied on nature. Ice in effect became a crop which was harvested in the winter and stored in ice houses until needed.

In mid-century Barking up to 3,000 people were employed in collecting and carting rough ice during the heavy frosts of January and February. The aim was to collect a sufficient amount to last the fleet until the following November. The ice house there, at Abbey Road, had a capacity of 10,000 tons.[11] Soon all major ports had at least one ice house—the first one at Grimsby had a thatched roof to help keep it cool. Hull and Grimsby's ice was also collected at first from ponds and pools in the surrounding countryside. This supply depended acutely on the vagaries of the English weather and a mild winter made it difficult to obtain an ample supply. Many ice companies were formed, usually by groups of smackowners and fish merchants, and as demand continued to grow they turned to Norway.

Some of the purest Norwegian ice exported to England was stored at the town of Droback which is on the fjord leading to Oslo. A few miles from Droback was a very pure lake fed by mountain streams and by 1865 this had been purchased by the Wenham Lake Ice Company who jealously guarded its waters from pollution. Of 44,823 tons exported from Norway in 1865 some 43,359 tons came from this source. The ice was 'reaped' by means of a sharp ice plough which divided it into 21-inch squares. Wedges were driven in to split up the blocks which were then gathered in large pine sheds. These had double walls, two feet apart, and the space between them was filled with sawdust. When the blocks were stored, sawdust had to be strewn between them to keep them from freezing into a solid mass. Ice, of course, is both a solid and fleeting commodity and about five per cent of its volume was lost through melting by the time it reached England.[12] Ice was usually shipped to England by the ice companies in schooners. Not all was destined for the fish trade. The Scarborough fish merchants and smackowners James Sellers and Henry Wyrill, for example, usually brought in sufficient ice to fill three specially-adapted warehouses, and as well as supplying the port's smacks they also satisfied domestic demand in the surrounding area.[13]

As voyages lengthened more ice was required and this proved expensive. In 1878 it was estimated that the yearly ice bill for a smack had risen to £100 which accounted for roughly nine per cent of a successful vessel's annual earnings. Moreover, as the smacks opened up distant-water grounds more and more time was spent sailing to and fro and less and less actually fishing. In the late 1870s it was reported that a Grimsby smack working alone on such grounds spent only one-sixth of its time trawling.[14] In order to improve efficiency a more intensive system of fishing seemed to be needed and the smackowners increasingly turned to fleeting.

Fleeting involved a group of smacks working together under the command of an admiral. Their catches were ferried by open boat to fast sailing cutters which took them into market each day. As a result, a smack could spend much more time actually fishing on distant grounds than one which went single boating. Fleeting had originated at Barking back in 1828 and by the mid-1830s the smacks trawling off the Isle of Man often worked in a similar manner, with one of them acting as a cutter to take their catches into Liverpool.[15] By 1864 Hewett's operated two fleets. Their Home Fleet generally worked the southern North Sea and the Short Blue Fleet worked in more northerly latitudes. The Home Fleet consisted of between 120 and 200 smacks and spent the spring off the Dutch Coast. From May to July the fleet worked from Terschelling to Ameland and in the autumn concentrated on the Dogger and Silver Pits.

Other major trawling ports adopted a similar system which was further refined and developed. When Hull smacks started going over to the Dutch coast around mid-century for the Easter fishing they had sometimes taken advantage of the sailing cutters servicing Hewett's and Morgan's fleets. Eventually, the Hull smackowners began to fit out their own cutters but they were not at first as successful as their Barking counterparts. The practice at that time was to pack the offal fish in large baskets of about sixteen stone and the prime fish in smaller baskets. These were then taken into Billingsgate by the cutters and the proceeds shared amongst the fleet. This procedure—known as bulking—worked well with Hewett's and Morgan's fleets, where a large number of smacks were in a few hands and crews were paid mainly by weekly wage, but at Grimsby and Hull where ownership was more dispersed and part of the crew worked on a share basis, it often produced great dissatisfaction amongst both smackowners and fishermen alike. It was felt that the lazy crew received the same reward as the hard workers and bulking took no account of the size or catching efficiency of the smack. At both Hull and Grimsby fleeting was originally limited to the spring and summer seasons and for the rest of the year the smacks went single boating.

In 1878 a new procedure was introduced which altered many a smackowner's view of fleeting. Instead of sending baskets of fish to market and dividing the proceeds collectively, each smack sent its fish to the cutter sorted and in boxes bearing its own name. They were then iced and carried swiftly to port. In future, each smack could be credited separately for the fruits of her own labour. This system had two obvious advantages over its predecessor. Firstly, it restored incentive, for hard work and good catches were clearly rewarded, and, secondly, boxed fish kept better than bulked fish, fetching higher prices on the market.[16]

Many Hull and Grimsby smackowners now found fleeting to be the more attractive proposition, especially as the yields from single boating seemed to be declining as voyages lengthened. Moreover, the introduction of steam cutters to service these fleets seemed likely to improve profitability even further. Hewett's had been the first to use steam cutters back in 1864 and Grimsby followed suit soon after boxing was introduced. At the beginning of 1878 iron steam-screw cutters were built for Grimsby owners in Middlesborough. Named *Precursor, Despatch* and *Celerity* they were 130 feet long, had a gross tonnage of 190 tons and were built for speed.[17] The following year another two craft were added. In January 1880 a meeting of Hull smackowners decided to form two fleets, the Red Cross and Great Northern, which would be serviced by four steam cutters apiece. The first two of these, *Europe* and *Asia*, fitted with sixty horsepower compound engines, were launched in the following April. All in all some £32,000 was invested in the venture.[18]

Steam cutters were now far more efficient than their sailing counterparts. When Hewett's had introduced them back in the 1860s it was found that they could reduce the average voyage from the fleet to Billingsgate in good weather to between thirty-six and forty hours[19] whilst the sailers might take between five and seven days. Steam-cutter fish had thus been fresher on landing, and when the sailing cutters followed them into port they found that their consignments fetched up to one-third less on the market. The new Hull and Grimsby craft boasted more efficient engines with lower fuel consumption and these cut transport costs considerably. The cost of conveying fish directly to London by Hull steam cutter in

1880 was reported to be 2s 2d per box, including porterage. If fish were carried into Hull and then forwarded to London by train, then the cost with ice and associated porterage was estimated to be 3s 8d per box. Bonuses were paid to key members of each cutter's crew on the number of voyages made to and from the grounds and it was estimated that four steamers could service more than 100 smacks whereas at least one sailing cutter was needed for every twenty.[20]

The introduction of steam cutters required a considerable capital outlay and smackowners had to form companies to finance their construction and operation. In order to recoup on this investment, and at the same time shore up the declining profitability of the smacks, they needed to operate the steamers throughout the year. However, many of those who manned the smacks viewed fleeting and the introduction of steam cutters from a rather different perspective. Most trawlermen preferred single boating, for fleeting meant spending up to eight weeks at a time on grey seas in all weathers on a cramped and wet smack with little vestige of home comfort. The idea of spending autumn and winter as well as spring and summer in such a fashion was viewed with little enthusiasm. Moreover, fleeting was considered a dangerous occupation. There was far more chance of vessels colliding when large numbers were working together and the actual task of ferrying boxes of fish from smack to cutter in an open boat was a most perilous occupation, made more hazardous by winter weather.

Despite such reservations on the part of the crews the smack-owners were determined to press ahead with extended fleeting. The first move at Grimsby was made in the autumn of 1878. When the fleet returned from the summer voyage, preparations were made to send the steam cutter *Celerity* out again with forty smacks. In the event she left on 12 September with about twenty smacks, and assurances of others to follow. Two days later a gale blew up and the fleet was scattered. Many smacks seem to have taken the opportunity to abandon the operation and returned to Grimsby a few days later claiming they had lost their Admiral. Other craft which had been due to join them changed their mind, took on board ice and went single boating instead. The Admiral, accom-panied by three or four other smacks, remained on the grounds and the *Celerity* continued to service them in the hope that others would join but the prejudice of the crews proved so great that the operation was wound up within a few weeks. All three cutters were then laid up in the docks until the following spring.[21] Nevertheless, many owners were determined that fleeting should succeed and pressed on with the construction of more steam cutters.

Attempts had been made to form a trawl fishermen's protection society at Grimsby in 1873 and again in 1876;[22] at the end of 1878 the men were trying once more to form a union with the aim of resisting the extension of fleeting.[23] The feeling against the use of steam cutters was by now so strong that in the following spring the police had to be called out to protect the first fleet of thirty-two vessels to set sail.[24] Faced with such opposition the owners let matters rest until the following winter when, in September 1880, the inevitable clash occurred.

The smackowners society resolved to carry on fleeting throughout autumn and winter, conceding only that the smacks should stay out for six weeks rather than the usual summer voyage of eight weeks. The fishermen's society, led by Joseph Cotterill, rejected all ideas of fleeting during the autumn and winter. Moreover, their determination was hardened by the so-called 'gum slip' issue; this was a notification slip which the owners stuck on the official Board of Trade engagement form and which raised the proportion of costs for nets, fish boxes and related gear that was to be deducted from the men's share of the proceeds of each voyage. Within a week some 250 fishermen had struck and more joined them as smacks came into port. Before long there were over 400 smacks lying idle in the docks and at least 700 men on strike. From the first, most of the men were reported to have conducted themselves with great dignity and they won the support of the townsfolk. By no means all of the fish merchants were behind the owners for, although cutters also brought cargoes into the Humber, many were worried about the amount of fish now being taken direct to London rather than through Grimsby. A fair few of the smaller owners, worried about their financial viability in a long struggle, were ready to break ranks. In contrast, the solidarity of the trawlermen held and within three weeks the dispute was over. At the end of September it was resolved by a committee of the Smackowners Association, after bitter debate, that individual owners should be relieved of their responsibility to the association 'with regard to the gum slip and should be at liberty to get their smacks to sea as they deemed best'.[25]

With no winter fleeting at Grimsby the Ice Company disposed of its steamers. Another Grimsby company made an unsuccessful attempt to introduce the practice again in 1885 and acquired five steam cutters to service its smacks. Once more a bitter dispute ensued. This strike, which the company was unable to break, was largely limited to about fifty or sixty smacks and dragged on through the autumn and winter.[26] Despite this opposition, modified steam cutters were eventually to play an important part in servicing the spring and summer fleets whilst finding employment as trawlers in winter.

The story at Hull was somewhat different. The trade there had some tradition of attempted unionization and had experienced industrial conflict. The first recorded strike in the Hull fishing industry had taken place in April 1852.[27] The dispute had arisen over the rules and conditions of service laid down by the owners through their Fisheries Protection League which the men viewed as tyrannical. The subsequent strike effort was concerted through the recently formed Fishermen's United Society, whose principal *raison d'etre* was to afford relief to widows and orphans. Both sides agreed to refer the dispute to the mayor for arbitration but when he expunged every rule in the smackowners' society and substituted new ones which favoured the men, the owners refused to be bound by his deliberations. They pressed on with the conflict and eventually broke the strike with the aid of smacks manned by apprentices. However, the concept of collective action amongst Hull trawlermen did not die for in October 1856 the owners, through their Fishing Smack Insurance Society, were again in conflict with an organization, then called the United Fishermen's Benevolent Association.[28]

The Grimsby dispute of 1880 was therefore watched with intense interest by all sides of the trade at Hull since the owners there also planned to introduce winter fleeting. They refused to recognize the Hull Trawlermen's Protection Association which soon claimed to represent over 400 men. The Grimsby men's victory increased their Hull counterparts' resolve to resist fleeting or at least gain additional remuneration for the greater risks involved and a large number struck from 18 October. The strike was organized by the union whose committee, chaired by C. H. Cook, met daily at its headquarters in Porter Street. Within a few days more men had joined and 150 of them paraded down the town's streets with a brass band and a banner inscribed 'God Defend the Right'. Most then attended a service in Fish Street Congregational Church.[29]

The strikers' hand was strengthened at the beginning of November when many smacks had to run for port at the same time to repair storm damage. By 8 November no less than 450 fishermen were reported to be on strike and over 100 smacks were laid up.[30] The owners, however, were strongly led by Owen Hellyer and the larger of them did not display that 'want of fidelity' which had been present amongst Grimsby's smaller owners. Many smacks continued to sail and the strikers did not win the level of financial support that they had hoped for. The year had been devoid of the sort of major tragedy that would have attracted widespread public sympathy. As the strike dragged on it became more bitter and a number of incidents were strongly dealt with by the magistrates. The strike committee helped those in greatest distress but their plight contrasted sharply with the high earnings of those who remained at sea. After eight weeks the strike was called off and the Hull smackowners had established the principle of winter fleeting.[31]

During the 1880s there were sometimes as many as twelve fishing fleets working in the North Sea from ports such as Yarmouth, Lowestoft, Brixham, Scarborough and Ramsgate, as well as Hull and Grimsby. Hewett's, who were now based in East Anglia, were the most experienced. The Scarborough owners, led by James Sellers, had formed their own fleet in 1880,[32] having previously joined those sailing from the Humber ports; but like many others they avoided fleeting in winter. Not all smacks went fleeting—even at Hull, Grimsby and Yarmouth a large number stuck to single boating. Those who did were part of a sophisticated and yet somewhat disorganized practice. The high point of the day was the arrival of the cutter and as soon as it was sighted each smack's open boat was loaded with fish boxes and rowed towards it with great speed. Then, as this 1885 eyewitness account shows:

> The vessel was broadside to the waves, rolling heavily . . . Around her pitched a multitude of boats, some jamming against her side, others fighting to gain this position of vantage and yet others unloaded and fighting to escape from the melee. The air was

> filled with voices of men shouting orders and expostulations, some in anger, some in jest. The confusion and uproar culminated around the hold mouth with boxes being heaped upside down, on their ends—anyhow—faster than they could be stowed away. When a boat arrived alongside one of its inmates, selecting the moment (when) the swollen mass of water lifted it to the level of the steamer's bulwarks, scrambled over before it sank beneath his dangling legs. A brief struggle and he was established on deck and now receiving boxes from his mate, the moments for doing so being selected as before. The man on deck having grabbed a 7 stone box half dropped and half threw it from its elevated position on the bulwarks to the deck. Now it was either dragged, pushed, thrown or kicked as near the mouth of the hold as the blockade would permit.[33]

The trawlermen's lot, when fleeting consisted largely of eight weeks of unremitting drudgery and danger, was governed by the towing and hauling of the trawl and the gutting and stowing of fish. Operations only ceased when storms made the seas too heavy or when the smacks were becalmed. Not surprisingly many crews sought relief from the 'bumboats'. These vessels, otherwise known as coopers, targetted the fishing grounds but had no intention of fishing. They sailed from ports such as Bremerhaven and Flushing stacked with cheap spirits, tobacco and other dubious wares which they plied amongst the fishing fleets. One vessel which had an English master named Bennington was forced into Grimsby in 1878 by foul weather, and when it was searched its stores not only included liquor, tobacco, coffee and tea but also forty-eight cases of perfumery and fifty-three packages of obscene playing cards.[34] Little money usually changed hands, for such goods were often bartered in return for fish or sometimes the smack's gear, which the men would later claim had been washed overboard. This practice incensed the owners, but as it took place in international waters there was often little they could do.

On returning to their continental bases the coopers swiftly disposed of these bartered goods. Indeed, one smackowner who visited several marine stores at Nieuve Diep in Holland, a regular base for coopers, reflected ruefully that they seemed to be full of fishing stores bartered from English crews. A number of the bumboat skippers were former English trawlermen who made the most of their knowledge of fishermen and fishing grounds. Many others were Dutch. The trade was highly profitable and one cooper was reported to have made as much as £900 in eight weeks from the fleets fishing out in the North Sea.[35]

Getting drunk on the cooper's liquor may have lightened life for a while but it also led to all manner of evil consequences. Back in the 1870s a Lowestoft smack master was convicted at Ipswich Assizes of piracy after a fracas with a Dutch cooper. He had initially purchased some spirits but on getting drunk had returned to the cooper with a number of others and forcibly taken its supplies. Drunken crews were dangerous and threatened both life and property. One typical incident occurred on the first day of spring 1880. That morning the master of the Grimsby smack *Cossack*, who was also a fleet Admiral, rowed across in the open boat to a Dutch cooper with his third mate and a deckhand. They returned with the master of the Hull smack *Britannia* and nine bottles of gin, rum and whisky, and by 3pm were well and truly drunk. The helmsman stuck to his duty but as he was passing two other smacks the master lurched on to the deck, struck him and then pushed the tiller hard over. The vessel then collided with the smack *Dawn* causing £40 worth of damage. By this time the master was in a blind rage and threatening to sink the smack. He threw the fishing gear overboard and let off all his signal rockets to confuse the fleet before rowing the *Britannia*'s master back to his vessel. Though the *Cossack* lay to for four hours he never showed any sign of returning and, as the other two drunkards lay insensible below, the two sober crew members set sail for home.[36]

On a number of occasions smacks were left in the control of a young apprentice whilst the rest of the crew went drinking on a cooper. There were instances when apprentices pawned their clothes for a bottle or two, and at least one, who was sent across to a bumboat for a supply of water, returned to his smack both naked and drunk. Now and again crews would slip into Continental ports where they illicitly sold their catch and used the proceeds to continue a drinking spree. Yet not all smackmen drank and a number who witnessed its worse excesses became strong supporters of the temperance movement. Others would use the periods of calm weather to gather on one vessel to sing, play music, hold religious services, or just talk.

Drink was far from the only source of conflict. The North Sea grounds were now crowded with Continental fishing vessels. The Belgians, Dutch and French in particular had developed their trawling trades after the manner of the English and the resultant congestion led to clashes. A prime source of friction was entanglement of gear. Trawlers had always been in trouble for catching up drift nets and the correct procedure in such situations was to fasten the drift warp on either side of where it was necessary to cut it with a line passed astern. This was known as 'knotting the warp' and caused only minimal damage. Not all British trawlers had adhered to this but now a number of Belgian smacks were accused of attaching a warp cutting device (nicknamed by the English a 'devil') to their gear. The position soon reached a critical pitch and in 1883 it was reported in the House of Commons that open warfare was carried on at sea with ballast stones and even firearms being used as weapons.[37] The Government had already commissioned a report and a captured 'devil' was exhibited in several ports. This situation forced the countries bordering the North Sea to cooperate on its policing and by 1893 they had ratified a number of conventions aimed at governing conduct at sea and controlling the liquor traffic.[38]

Few people knew much of the rough and brutal life of this neglected navy of seafarers, but one, who visited them at work determined to improve their spiritual and bodily welfare, was Ebenezer Mather, a Londoner who worked in the Thames Church Mission. His work brought him into contact with many merchant seamen but few fishermen and in 1881 he decided to visit them at work on the fishing grounds. Mather made the voyage to Hewett's Short Blue Fleet in the steam cutter *Supply* and his brief investigation both amazed and appalled him. He returned to port fired with the desire to make their life more tolerable and founded the Mission to Deep Sea Fishermen which soon became both Royal and National. This mission was to go amongst the fishing fleets, and he raised money for the venture by campaigning energetically amongst smackowners and the general public. The first mission vessel, a converted smack called *Ensign*, sailed for the grounds staffed by a minister and medical dresser—previously no medical help had been available. A fleet of specialist mission craft was soon acquired and this provided the fishermen with medical aid, warm clothing and baccy as well as bibles—but no booze. The mission smacks were kept busy and their services were well attended by fishermen.[39] Their efforts were soon greatly appreciated by those who had previously relied on the dubious comforts of the bumboat. The Royal National Mission to Deep Sea Fishermen still carries on its good works today.

The issue of winter fleeting sprang into prominence once more in 1883 when the Hull trawlermen again sought to end it. Their determination was hardened by disaster. On 6 March 1883 a sudden and severe storm had broken upon the hundreds of fishing vessels out in the North Sea. All in all, some 250 men were lost along with forty-three smacks; countless others returned to port in severely damaged craft to recount terrifying tales. Hull, Grimsby and Scarborough vessels were hit worst. Indeed, 180 of those men lost hailed from Hull alone. They had been working along the northern edge of the Dogger where the north to north-west wind had whipped up a sea of unusual size and violence. The Yarmouth fleets had been working south of the Dogger—Hewett's were on the Swarte Bank—and suffered less.[40] The sheer scale of this disaster, and the plight of the widows and orphans left behind, attracted an enormous amount of public sympathy for the fishermen's cause. That summer the Hull Trawl Fishermen's Society—which claimed to have used a third of its funds on funeral expenses after the March gale[41]—determined to end winter fleeting and gave notice to the smackowners that sailings would cease on Monday 1 October 1883 if their demands were not met. They requested that the steam cutters be laid up and the smacks fitted out for a winter's single boating.

Though there was little animosity at first, bitterness grew on both sides as the owners continued to try and get smacks away to sea. The owners alleged that the strikers were trying to stop all vessels, whether fleeting or single boating, but the union claimed that they were letting skippers sail who were working off a mortgage and would not stop any owner sailing out on his own boat. The owners protested to the town authorities that they were being given insufficient protection to maintain the liberty of their property. A few days later four men, including James Carrick, president of the fishermen's association, were arrested and charged with obstruction when a crowd tried to prevent police seeing a man safely on to his smack. The subsequent case, something of a local *cause célèbre*, lasted two days and in the end was dismissed. Afterwards, Carrick was carried around the town in triumph by the jubilant strikers.[42]

The strike held into November though many families were suffering great hardship. The owners were equally resolved to hold out and the dispute seemed set to last awhile. It was finally brought to an end on 12 November 1883 through the mediation of Hull Trades Council. The result was a compromise. The men failed to secure the total ban on winter fleeting but the owners agreed to limit winter fleets to a maximum of sixty vessels—which was felt would reduce the risk of collisions—and not to work them on the dangerous grounds north of fifty-five degrees.[43] Henceforward, less smacks went fleeting in winter. Hellyer, one of the first to be involved in steam cutters sold off his interests and worked his smacks on the single-boating system. By 1886 it was reported that up to four steam cutters were being laid up for the winter because of the smaller fleets.[44] Yet disputes continued to afflict the trade nationally during the 1880s. In the spring of 1887, for example, there was a strike amongst Scarborough trawlermen over reductions in their share of vessel earnings and they forced the owners into partial concessions.[45] But such victories did nothing to alleviate the harsh economic fact that smacks were becoming less profitable as catches and thus earnings declined. During the 1890s most ports gradually abandoned fleeting and by the early twentieth century Hull was left as the only port continuing the practice.

The series of controversies which afflicted the trade during the early 1880s had left their mark. The Government had been obliged to conduct a number of inquiries and enact legislation covering a range of matters including conditions of engagement for apprentices and crew, vessel lighting, and other safety measures. Much of this was incorporated into the 1883 Fishing Boats Act which, amongst other things, required both masters and mates of vessels over twenty-five tons to be certified, and obliged each skipper to keep a log recording every case of death, injury and ill treatment etc. No youngster was allowed to be signed up as crew under the age of sixteen years unless he had been properly indentured as an apprentice. Government 'interference' in the trade's affairs had returned.

7

The Coming of the Steam Trawler

Until the late 1870s the British trawling trade relied almost completely on sailing vessels for catching fish. The dramatic rise of the 'new' fishing ports of Hull and Grimsby was based upon the creation of large fleets of smacks which ranged across the North Sea grounds. Their investment in these craft had yielded substantial benefits. The trade at Hull had been originally limited to an area of river front near Nelson Street by the ferryboat pier and although it had soon moved into Prince's Dock, it rapidly outgrew available facilities. In 1869 the smackowners were permitted to use a portion of the newly opened 24½-acre Albert Dock but in 1883 they were to secure exclusive usage of the freshly constructed St Andrew's Dock which remained their base until the 1970s. There were some 420 smacks sailing out of the port by 1882 and these supported a whole range of ancillary trades. Many had been constructed in local yards and provided employment for numerous blacksmiths, shipwrights, sail-makers and riggers. Large numbers of people found work in the carting and processing of fish and over fifty smokehouses had been constructed. By the end of 1883 the Hull fish trade's capital was estimated to be worth more than £1 million and provided direct or indirect employment for more than 20,000 people.[1]

Developments at Grimsby had been even more impressive. Since the opening of the first fish dock in 1856 there had been a massive expansion of the trade. The smacks had twenty-one acres of water and quay after the second fish dock was completed in 1877 but this was soon considered an inadequate area, for by 1880 the port could boast a registered fleet of 567 first-class smacks. That year the railway companies' revenue for fish forwarded by train exceeded £100,000 whilst a greater amount was ferried direct from the fishing grounds to London's Billingsgate. As at Hull, a wide range of ancillary trades had grown up, including two ice companies and the Coal, Salt and Tanning Company. This latter enterprise had a paid up capital of over £8,000 and tanned the sails of smacks and also supplied them with coal, salt and paint etc. A substantial labour force was needed to build and maintain the smacks as well as deal with the fish they caught. By 1877 the trade required more than 250 ship's carpenters and 100 sail- makers, numerous smiths and a range of related tradesmen. There were more than fifty coopers and over 400 men, women and boys worked in the forty-one smokehouses.[2] Many others processed and moved fish on the quayside or found employment in the manure works. Hundreds were kept busy in roperies or making nets. Here also, over 20,000 were said to be dependent on the harvest of the smacks and the same story was true on a lesser scale of all other trawling ports up and down the coast.

Yet over a relatively short space of time the sailing smack was displaced by steam. In 1876 no commercial steam trawlers were working from British ports but over the next five years a large number of paddle tugs were adapted for fishing. Shortly afterwards, purpose-built steam-screw trawlers began to appear. By the end of the 1880s Hull and Grimsby had embarked upon a large-scale replacement of smacks by steam trawlers whilst the new trawling centres of North Shields and Aberdeen based their development almost completely on steam.

The Victorian fish trade, has often been criticized for being slow to take advantage of steam[3] but more recent research has shown that the importance of steam during Britain's emergence as the first industrial nation has been over-emphasized.[4] Many trades continued, for rational economic reasons, to rely on more traditional sources of power, including wind and water, well into the second half of the nineteenth century. Neither can it be argued that the introduction of steam was delayed by any innate conservatism afflicting the trade. Though there were certainly fishing communities reluctant to alter traditional custom and practice, there were strong and progressive elements at several trawling ports eager to adopt economic innovations, as the spread of beam-trawling, the introduction of ice and the continuous development of the sailing smack clearly shows. There had also been a number of attempts to apply steam power to the catching sector before the last quarter of the nineteenth century.

Marine steam power had been demonstrated as an economic proposition on some trading routes over the thirty years following the introduction of steam boats on the Clyde in 1812. Yet until the 1860s, when the simple single-cylinder engines and low-pressure boilers—ravenous consumers of coal—began to be displaced by more efficient and powerful compound units with high-pressure boilers, the steam vessels's relative inefficiency limited its commercial application.

By the 1850s experimentation with steam fishing vessels had begun. In 1853 a steamer had been built for the Deep Sea Fishing Association of Scotland[5] and several Grimsby steam fishing vessels had entered service circa 1854/6. These ventures proved unsuccessful and the Grimsby vessels had their engines removed after failing to cover operating costs. In the 1860s several unsuccessful attempts were made to operate steam paddle trawlers, including one in the approaches to the Firth of Forth and another in Waterford Bay, Ireland.[6] Another attempt to fit engines in an iron-hulled Grimsby smack in 1870 proved a failure. In 1867 George Bidder, an esteemed engineer, was involved in the construction of several steam trawlers at Dartmouth which proved commercially unsuccessful in the longer term. Several steam fishing vessels were constructed by Scottish shipbuilders for French owners in the 1860s. Scottish shipbuilders also constructed a steam fishing vessel called *Onward* in 1877 about the time steam trawling was proving successful in England.[7]

The fish trade did apply steam successfully in a less direct manner. Steam packets and steam trains were used for forwarding fish to market and, of course, during the 1860s Hewett's of Yarmouth pioneered the use of steam cutters for carrying fish from the fleets direct to London's Billingsgate. Other sailing fishing vessels often employed steam tugs to tow them to and from grounds when unhelpful weather conditions prevailed. Another, though somewhat later, innovation adopted after 1876 was the steam capstan which eliminated the back-breaking job of hand-hauling the trawl.

It is thus apparent that the fish trade found a number of ways of using steam whilst remaining reliant on sail in the catching sector throughout the third-quarter of the nineteenth century. The role of the steam engine in the fish trade during the 1860s and 1870s can be compared in certain respects to that of the mainframe computer a hundred years later. The latter, of course, was both large and expensive and this often precluded its general use for small-scale operations. It tended to prove commercially viable only where activities could be organized on a relatively large scale allowing maximum use to be made of the capital invested. If and when small-scale organizations wanted to make use of mainframe computers then they normally rented or bought processing time. Widespread computerization did not take place until the silicon chip and the microcomputer were introduced. Similar economic forces can be seen operating in the fish trade of the 1860s. A basic paddle steamer with a simple low-pressure engine cost around £3,000, or more than three times the cost of a contemporary sailing smack. The first compound engines were more economical when fitted to somewhat larger craft but as fishing vessels still worked comparatively close to home for most of the year the swifter steamer had only a limited advantage in terms of speed. Furthermore, steamers required a larger and more skilled crew whilst smacks could rely to a considerable extent on cheap apprentice labour. Thus the smacks merely hired a tow from steamers when conditions required or were only serviced by a steam cutter when large-scale fleeting in more distant waters was undertaken.

The first steam vessels to make a commercial success of fishing were converted paddle tugs. According to legend some sailing smacks dropped their trawls whilst being towed by the paddlers and from this it was only a small step to trawling direct from the tug. In 1877 a severe trade depression afflicted the ports of north-east England and the steam tugs were early casualties. The drop in trade meant less ships entered the Tyne and Wear and this reduced the towing work available for the local tug fleets. Their business was already suffering from the growth of steam shipping fleets which required less support. Scores of tug boats were soon laid up along the Tyne with little obvious prospect of employment.[8]

In an attempt to keep his vessel paying its way, one North Shields tug master, William Purdy, decided to try trawl fishing. He fitted out his ancient steam tug *Messenger* by adapting whatever gear was available locally but had to send down to Grimsby for a beam trawl. His first trawling trip was made in November 1877. In many ways the port was ripe for exploitation by the trawling trade. Back in 1866 North Shields Corporation, keen to promote the growing herring trade, had embarked in a modest way on the construction of a fish quay. During the following decade it expended a great deal of capital on purchasing land for the trade at Low Lights, and in reconstructing and extending the quay facilities. With the herring trade being a seasonal affair the Corporation Fish Quay, as it was already known, was under-used when Purdy sailed off.

Initially, Purdy's venture was scoffed at but, although the earnings from his first voyage were only modest at £7 10s, his gamble soon paid dividends.[9] Though the *Messenger* and similar paddle steamers were often quite elderly and had only simple side-lever engines, their fuel costs were drastically cut by a long-term fall in coal prices which was another feature of the trade depression. By mid-December some fifteen other North Shields steam tugs had followed Purdy's lead and the number had reached forty-three by the end of the winter.[10]

At first, many of these paddlers opted for a dual purpose existence: fishing in winter and returning to towing when demand picked up during the summer months. Only eighteen were reported to have continued trawling out of North Shields over the summer of 1878. Because their skippers lacked fishing skills they had to employ experienced hands on deck. Quite often the tug skipper, who had overall charge of the boat and navigation, was accompanied by a fishing master, who looked after trawling. Soon the fishing interest was buying its way into the fleet and a number of men began to acquire the all-round skills necessary for trawling under steam. North Shields benefited quickly from steam trawling and it was soon necessary to embark on an extension to the Corporation Fish Quay. Purdy's colleagues were so grateful that in April 1878 they held a supper in his honour at the town's Sun Inn where they presented him with a handsome testimonial consisting of a silver lever stopwatch and an inscribed gold medal.[11]

More and more of these paddlers found it worthwhile to operate throughout the year and by the winter of 1878/9 some fifty-three were working out of Shields. The innovation caught on at other ports. By late 1878 six or seven were fishing out of the Wear and others could be found at Hartlepool. As early as January 1878 the Grimsby steam tug *United* experimented with trawling and was able to supplement her earnings from her first trip by towing home a distressed barque.[12]

Unlike the smacks, which by now concentrated on offshore grounds, the converted tugs usually worked within ten or twelve miles of land. They clung to the shore more out of necessity than choice for, being designed mainly for towing work in the vicinity of estuaries and harbours, they were not built with great sea-keeping qualities. This obliged them to run for shelter whenever the weather deteriorated. If caught, the consequences could be dire and many were lost including the *Nation's Hope* of North Shields which went down with all hands off Port Mulgrave during a great storm in October 1880. They were also heavy on fuel and, being built for towing, storage space was at a premium. What room there was had to be shared between catch and coal. Quite often the latter could only be carried by cutting bunker capacity, reducing their already limited range.

Given these shortcomings the most profitable way to operate was to restrict voyages to less than twenty-four hours and to fish as intensively and continuously as possible. Coal was not wasted steaming to and from distant grounds, whilst shelter was usually close by. By having their trawl down as much as possible they were able to make up for times when bad weather confined them to port and, being paddle steamers, their excellent manoeuvrability enabled them to work small pockets of smooth ground close to the shore that the smacks had found virtually impossible to trawl.

Though working inshore, the north-eastern steam paddlers did not just trawl from their home ports but began wandering quite far afield. Some tried their luck off the Scottish coast and before the end of 1878 they were occasionally landing at Scarborough. Within a couple of years more than a dozen north-eastern steam trawlers might sometimes be observed landing fish in Scarborough Bay.[13] By working close to the shore they were in direct competition with Yorkshire and north-eastern line fishermen who—no longer troubled so much by smacks—bitterly resented these new incursions on to what they regarded as their traditional grounds. Though the

Plate 2. The Scarborough Steam Paddle Trawler Fearless *Moving Astern off Scarborough Harbour* c. *1890*

smacks worked further out they still had to compete when landing at the same quayside markets. Indeed, when the steam paddlers chose to concentrate on Scarborough their sheer numbers, combined with their catching power, could flood the market; moreover, being steam powered they could usually land before the smacks and obtain the best prices. As they returned to port each day their fish was generally fresher than that caught by smacks returning from a voyage of a week or ten days.

The Scarborough trade was initially reticent in opting for paddle trawlers. Nevertheless, the port's first steam trawler, *Cormorant*, had commenced fishing in the summer of 1878.[14] This 62-foot long craft differed from the Tynesiders in that she was not only screw-driven but primarily a pleasure yacht. Built by Messrs Richard Smith of Preston and fitted with their own direct-acting compound steam engines, her owner was Henry Hird Foster, a wealthy Scarborough gentleman who employed a local yawl fisherman,

William Appleby, as skipper.[15] *Cormorant* towed a beam trawl 33½ feet wide and 46 feet in length.

By the end of 1880 it was noted at Scarborough that the north-eastern paddlers 'bid fair to gain a monopoly of the trade' and at least two were reported to have picked up £70 on the fish market there from just 20 hours fishing. The Scarborough traders were soon following suit and their first paddler, *Dandy*, arrived before the end of December 1880. Like many of the others she was no youngster, having been built at Willingham Quay on the Tyne back in 1863, and she had seen service as a tug at several ports including Dublin and Liverpool. The *Dandy*'s new career got off to an inauspicious start however, for a crew member, James Field, was killed when a chain snapped as the trawl was hauled up, and on returning to harbour she collided with the pier and damaged a paddle box.[16] Nevertheless she soon proved successful and by the end of 1881 there were ten paddle trawlers registered at Scarborough.

Paddle tugs were by no means cheap to acquire and even a quite elderly example might cost about £2,000. Once their potential as trawlers was realized the asking price often rose nearer to £3,000, or twice the price of the latest sailing smack.[17] Such capital could be difficult to raise and even stretched many of Scarborough's well-established smackowners and fish merchants. Though a few were bought outright by individuals, most were acquired by newly formed private and then limited liability companies. Amongst the earliest of the latter was the Yorkshire Steam Trawling Company Limited which was established late in 1881 and which soon operated several vessels. The smackowners and fish merchants who formed the backbone of such enterprises had little trouble mobilizing capital from outside the fish trade and one company's £25 shares were almost immediately changing hands at a premium of 30s.

The Scarborough paddle tugs appeared a sound investment on their performance during 1881. They landed good catches on favourable markets and met only a minimum of bad weather. Their apparent success encouraged a somewhat over-optimistic assessment of their future potential and led to a fashion for paddle steamers in 1882. During the three months ending 28 February 1882 a further ten steam fishing vessels were acquired and the fleet reached its zenith in 1883 when some twenty-seven steamers could be mustered.

Inevitably this period of buoyant optimism fostered several less-than-sound ventures and amongst these was the Knight of the Cross Steam Trawling Company. This company raised just over £2,391 to purchase a steam tug of the same name from Liverpool in January 1882. *Knight of the Cross* was the largest and most powerful member of the Scarborough fleet, possessing a 121-foot long hull and a vertical side-lever engine rated at seventy horsepower, which gave her a burning appetite for coal. Like several other Scarborough paddlers she was expected to supplement her trawling activities by running summer passenger trips along the coast. The company's backers confidently asserted that she would be able to fish as far afield as the Norwegian coast and, with a full catch on board, could earn £300 on a good market. The twenty-year-old craft was, however, evidently in poor mechanical condition and regularly laid up for repairs. Her income could not match the expenses she incurred. In late January 1883 the company's shareholders decided to cut their losses and wind it up. The venture had been nothing short of a financial disaster,[18] and unfortunately for the Scarborough fishing community the company was not the only firm to go down.

1883 proved a difficult year for the steam paddlers. They could only remain profitable if worked intensively, but adverse weather conditions regularly confined them to harbour which considerably reduced their earnings. Moreover, paddle trawlers had been working the inshore north-eastern grounds almost continuously for over five years with an ever-increasing intensity and, not unnaturally, these grounds began to show signs of exhaustion. This not only lowered catches and earnings for the paddlers but it had dire consequences for the inshore men. The former at least had the opportunity to voyage down to unaffected areas of coast whereas the inshore fishermen's mobility was greatly restricted. Even so, this lack of regard for the industry's resource base, coupled to the bad weather, resulted in a series of steam trawling company failures at Scarborough, the worst affected port. In February 1884, the Yorkshire Steam Trawling Company, already deeply in debt to the bank, was wound up and several other ventures went the same way, including the Star o'Tay Steam Trawling Company. Other sectors of fishing were inevitably affected by the poor catches and several leading Scarborough entrepreneurs went bankrupt over the following three years.[19]

Gradually, all ports lost interest in the paddle steamer phenomenon though Scarborough persevered much longer than most. Throughout the 1890s the trade there maintained a fleet of up to ten such craft but because the local grounds were so denuded they often had to work on inshore grounds off distant coasts. In 1897, for example, Scarborough paddle trawlers were reported to be working in the Irish Sea and landing their catches at Milford Haven.[20] A few even tried their hand off the west coast of Ireland. The paddle trawler chapter did not finally close until Scarborough's last example, *Constance* was wrecked at Hartlepool on 22 March 1910.

Though Hull and Grimsby never took up the paddle steamer mania they were soon to assume a major role in the development of the purpose-built steam-screw trawler. Initially, attempts were made to add the steamer's speed and manoeuvrability to the seagoing qualities associated with traditional smack design. Towards the end of 1880, for example, a Scarborough boatbuilder, John Edmond, started to modify a conventional smack he was constructing so that it could take a steam engine. This vessel, the *Young Squire*, was fitted with a diminutive six-year-old ten horsepower steam engine and boiler built by Plenty and Sons of Newbury, Berkshire. However, this original engine seems to have been underpowered for in the autumn of 1881 she was refitted with larger twenty horsepower engines supplied by the same engineers.

Whereas the *Young Squire* was basically an auxiliary powered sailing smack, the next steam-screw vessel to be registered for fishing at Scarborough could perhaps be called the first purpose-built steam trawler. The *Pioneer*, registered at Scarborough in October 1881, was an iron-hulled craft and, at ninety-four feet in length, was far larger than any conventional sailing smack. *Pioneer* was built by John Shuttleworth of Hull and fitted with thirty-five horsepower direct acting engines supplied by Messrs Pattison and Atkinson of Newcastle upon Tyne. Though nine individuals held shares in the trawler, her prime owner and mover was James Sellers.[21]

Before the end of the year Grimsby had also acquired its first purpose-built steam-screw trawlers. During August a group of leading smackowners and fish merchants, amongst them H. Mudd, J. Alward and C. Jeffs joined forces with several individuals connected with the Manchester, Sheffield and Lincolnshire Railway to form the Grimsby and North Sea Steam Trawling Company. This concern, which had an authorized capital of £50,000 in £10 shares, soon had two steam trawlers under construction. The *Zodiac* and *Aries*, built by Earles of Hull and T. Charlton of Grimsby respectively, were launched in mid-December 1881. The *Zodiac* was some ninety-four feet in length and, being fitted with thirty-five horsepower compound surface condensing engines, could reach ten knots.[22] Neither craft dispensed entirely with the wind for their engines were disconnecting so that they could proceed under sail alone if necessary. Both were also able to work as carriers for the boxing fleets in summer. *Zodiac* and *Aries* proved so successful that the company ordered four more the following year. Other steam-screw trawlers were soon being built but their progress at Grimsby during the 1880s could best be described as steady rather than spectacular. By 1890 Grimsby could boast a fleet of forty-two steam trawlers.[23]

Hull was slower getting started, though shipbuilders there had turned out at least seven steam trawlers for other ports by the end of 1885. Though there had been one earlier attempt, steam trawling did not take off in Hull until the latter half of 1885. At first, steam carriers, which were now under-used thanks to the decline of winter fleeting, were adapted but purpose-built vessels soon followed. The port's pioneers included the three firms of G. Beeching, Pickering and Haldane and the Hull Steam Fishing and Ice Company. The establishment of steam fishing was assisted by the wealth of marine engineering expertise available in the port from companies like C. D. Holmes, and over the following years many vessels were to be turned out by such firms as Earles and the newly formed Cook, Welton and Gemmell. Cook and Welton had been platers at Earles whilst Gemmell had worked for the same firm as a naval architect. Their first vessel had been an iron smack built for Robert Hellyer in 1885 but by the end of the year their newly built steam trawler *Irrawaddy* was having trials for its owner George Beeching. By the time Cooke, Welton and Gemmell moved to a new yard up the River Hull at Beverley at the turn of the century, they had built around 350, mostly steam powered, trawlers.

As at other ports, the conversion to steam and its associated high demands for capital encouraged the creation of large-scale limited liability companies. In 1886 the Humber Steam Trawling Company was formed with an authorized capital of £30,000 (raised to £100,000 in 1891) and amongst the others was the British Steam Trawling Company Limited founded in 1889 with an authorized capital of £50,000 (raised to £100,000 in 1891). Moreover, several leading private firms also converted to limited liability, including Pickering and Haldane in 1889 and C. Hellyer in 1891.[24] Similar companies were, of course, formed at other ports and proved a successful means of attracting and mobilizing capital from outside the fish trade on an unprecedented scale. By 1890 North Shields, with eighty-one vessels, could boast the largest fleet of steam trawlers in the world; Hull, with sixty-one, had the second largest, and possessed more steam-screw trawlers than its Tyneside rival, whilst its fleet of smacks was still nearly seven times as large.

North of the border, trawling had hitherto made little impact. There had been occasional attempts to establish the practice—steam trawling had been tried in the approaches to the Firth of Forth as early as 1863—but few Scottish fishermen were interested in taking it up. True, a few sailing craft from one or two east-coast stations fitted out each year for seasonal trawling but most Scotsmen remained contemptuous of the practice and bitterly resented what they regarded as incursions by English boats on their traditional fishing grounds.[25]

Yet the activities of these English trawlers were soon to encourage the remarkable development of Aberdeen as a trawling port. Prior to the 1880s Aberdeen's chief involvement with the fish trade had been as a base for the visiting herring boats and curers during the summer season. A handful of small inshore lining craft worked out of the neighbouring communities of Torry and Footdee and their catches came to be supplemented by the occasional landings of English trawlers working in the area. As English trawlers started taking advantage of the good catches that were then being made in Aberdeen Bay and on grounds further north, it was natural for them to offload their full holds at the port on almost a daily basis rather than return regularly to their home ports. Facilities were rapidly developed to deal with this new trade and daily auctions introduced in 1881.

No trawlers were recorded as being based at Aberdeen when the port's register of fishing vessels was opened in 1869 but in late

1875 two eleven-ton sloop-rigged vessels, the *Ann & Elizabeth* and the *Ocean Mail*, were described as fishing with trawl nets when entered on the register.[26] From late 1879 a small but increasing number of inshore sailing vessels took up trawling in the neighbourhood and George Brown of Summer Lane registered his Grimsby-built smack, *Ruth*, there in February 1881.[27] The port's first paddle trawler, *Toiler*, arrived in March 1882,[28] and during the winter of 1882/3 the Aberdeen fish trade acquired two steam-screw vessels, *Bonito* and *Kingfisher*, the latter of which had been built at Granton.

A further eleven more steamers were entered on the Custom House books during the rest of 1883 and Aberdeen's own steam trawling fleet was firmly established. The majority were converted paddle steamers after the North Shields fashion but one of the arrivals was the first purpose-built steam trawler, *Pioneer*, which was acquired from Scarborough. Few new vessels were registered at the port over the next few years as the steam trawler mania died away and the trade there continued to rely heavily on landings from visiting vessels, but from the middle of 1888 there was a great surge in new registrations of purpose-built steamers thanks to the initiatives of William Pyper and others.[29]

By 1890 Aberdeen had thirty-eight steam fishing vessels and most were trawlers.[30] Their efforts continued to be supplemented by landings made by visiting trawlers,[31] and landings of fish other than herring had risen from almost negligible proportions to over 250,000 cwt in 1890. The expanding steam trawling trade attracted fishermen and merchants from far and wide. Many soon made it their home. One typical arrival was Charles Winton whose story illustrated the development of the national trawling trade over the previous forty years or so. Winton was born in Brixham and had gone to sea on the smacks in the 1840s. He had trawled off coasts from Lands End to John o' Groats; he had sailed from Hull, Yarmouth, Sunderland and Newcastle, and by 1883 was on steam trawlers based at Aberdeen.[32] Indeed, many of those recruited as crew on the port's first steam trawlers came from England. Lots of Scarborough trawlermen were to migrate northwards, for example, and as late as 1900 at least one Aberdeen trawler was still manned entirely by Scarborough men.[33]

Distance restricted Aberdeen's penetration of English markets and although a London-bound fish train service had been introduced in 1883 the high rail rates at first limited this service to about one or two wagons. Nevertheless, a considerable trade was built up both locally and in the growing urban areas of south-west Scotland with a great emphasis being placed on cured fish. The region had long been noted for its smoked haddock, and Aberdeen fish market soon experienced heavy demand for haddocks which were cured in an ever-growing number of smokehouses.[34] A great deal of dried cod was also turned out but instead of being dried by sun and wind in the traditional fashion on open beaches much of it was laid out on racks and subjected to artificial draught and heat.

Aberdeen had quickly started building its own steam trawlers and local shipyards soon developed a considerable level of expertise, building up a substantial trade in the construction of fishing vessels for other ports in Britain and abroad. Steam trawling had also taken root at Granton and Dundee. Although those ports were able initially to keep pace they were soon left behind. By the early 1890s Aberdeen was clearly Scotland's greatest white fishing port, accounting for twenty per cent of the total catch.[35] By 1913 the proportion reached seventy per cent and the port boasted a fleet of 218 steam trawlers.

Back on the Humber, the pressure on existing facilities encouraged the fish trade to look further afield than Hull and Grimsby and in 1885 the Boston Deep Sea Fishing Company was formed with the intention of making Boston, the ancient Lincolnshire port, an important trawling centre. Soon Earles of Hull were building a fleet of eight vessels[36] and the leading Hull smackowner and fish merchant, Alfred Wheatley Ansell, was recruited as first managing director on a salary of £800 per year. The company's early years proved far from easy. It parted company with Ansell in an acrimonious fashion and a court action which it brought against him to recover damages found in his favour.[37] Though Boston offered less crowded port facilities it proved hard to recruit experienced trawlermen there, and navigation up the channel was also difficult. Eventually, the company moved to the Humber where it not only survived but grew into one of the most important of the twentieth-century trawling firms.

The 1880s witnessed marked improvements in marine steam engine design. In 1884 the triple expansion engine was introduced and boiler pressures continued to be increased. By the end of 1887 boilers were being built with a maximum pressure of over 150 lb and before long the 200 lb barrier was passed.[38] Only a fraction more than a pound of coal per horsepower per hour was used where the old low-pressure marine steam engines required ten. This drastic lowering of fuel consumption gave all varieties of steam ships a greater range for a given capacity of bunker fuel.

As trawlermen gained expertise fishing with steam, many of these developments were quickly taken up and steam trawlers were soon no longer constructed with additional sail. In broad outline, if not basic size and power, steam vessels such as the trawler *Dalhousie* of Scarborough, constructed in Dundee in 1886 and fitted with a thirty-eight horsepower triple expansion engine, differed little from the steam trawlers being constructed in the 1950s. The steam trawler's basic design evolved in less than ten years and at many ports smacks were soon sailing into history.

8

Beyond the North Sea

After 1873 the British economy steadily entered a period which became known as the Great Depression. It was characterized by a sustained fall in the price of most goods, whether domestic or imported, and lasted until the mid-1890s. The British primary producers—that is those supplying foodstuffs and raw materials—generally experienced the greatest decline in prices. Yet it is really inaccurate to characterize those years as ones of depression for the British economy as a whole. Though prices of manufactured goods fell, producers found that the prices of their raw materials were falling faster. Wages fell, but less drastically than prices and those in work actually experienced a growth in their purchasing power and thus standard of living. Moreover, it has been shown that Britain's overall economic performance, in terms of output and growth, was far better than was once supposed.

The causes of the decline in prices were tied closely to the development of the international economy. The opening up of resource-rich frontier lands and improvements in transportation systems, allied to migration and a growth in overseas investment by Britain and her European neighbours, brought more foodstuffs and raw materials on to the market thus lowering their cost. This had a knock-on effect throughout the whole economy.

Such international restructuring affected the pattern of economic activity. It made some activities less profitable whilst having the opposite effect on others. Nowhere was this more evident than with agriculture. Massive imports of grain from the North American prairies and Russian heartlands, courtesy of new railway lines and steamships, led to the ruin of many British cereal producers. But as staple foodstuffs including bread fell in price, other farmers, engaged in the production of dairy and meat products, benefited from the rise in real incomes which enabled the consumers to buy more of their products.[1]

The fish trade could scarcely expect to emerge unscathed from such a radical restructuring of the economics of food production. At first sight it might appear that such changes could account for the falling profitability of the trawling smacks which was increasingly commented upon from the later 1870s. Given the traditional view that fish was generally less favoured, and even regarded as a poor substitute for meat, then a rise in real incomes might have been expected to cause a shift in preference from the former to the latter. The Scottish herring trade certainly experienced a dramatic slump following a collapse of prices in 1883. However, this seems to have been brought about by over-production for the continental export market.[2] Though only a few price statistics have been unearthed for these years, they show, contrary to expectation, that, far from falling, many varieties of white fish actually experienced considerable price rises during this period.[3]

With the benefit of hindsight, including greater access to statistics and years of related research, it seems that the diminution in fish stocks on traditional North Sea trawling grounds was a major cause of the trade's problems. Yet for many sections of the contemporary trawling trade this was difficult to accept for a considerable time. Fishing, by its very nature, was subject to periodic fluctuations in catch size and the famous 1866 Royal Commission chaired by Professor Huxley had, of course, concluded that man's efforts could not affect the level of fish stocks.[4] Huxley's views carried considerable weight. In a widely publicized lecture given to the 1883 International Fisheries Exhibition, he reiterated his belief that:

> the cod fishery, the herring fishery, the pilchard fishery, and probably all the great sea fisheries are inexhaustible; that is to say that nothing we do seriously affects the number of fish. And any attempt to regulate these fisheries seems consequently, from the nature of the case, to be useless.[5]

The lack of available statistical evidence—the state did not really commit itself to any national system of data collection until 1886—further confused the picture. Furthermore, since at least the passage of the 1868 Sea Fisheries Act, the trawling trade had enjoyed being able to fish with the minimum of legislative restriction and there was an in-built reluctance to accept the need for state interference in the pursuit of conservation. Yet later research by Garstang at Grimsby and other places, from the limited statistics available, indicated that overall landings per smack had declined markedly between the mid-1870s and the early 1890s.[6] Garstang's findings were based on only a limited data source—four Grimsby smacks—but seems to back up many of the trade's later assertions. Falling yields were at first countered by moving ever further afield to open up newer or less exploited grounds and by adopting more intensive methods of fishing, including the construction of larger smacks and the widespread adoption of fleeting, as well as by the rise in the value of certain varieties of fish.

Such measures may have worked in the shorter term but they did not address the fundamental problem of stock diminution. At best they merely slowed rather than halted the declining profitability of the smacks. The problem of stock diminution off the north-east coast in particular seems to have been further aggravated by the introduction of the steam trawler, especially the North Shields and Scarborough paddle tugs, which had worked on the inshore grounds with an ever-increasing intensity since late 1877.

Steam trawling became an especially acrimonious issue along the Scottish east coast after 1882. The local line fishermen were incensed by what they viewed as unwarranted incursions onto their traditional grounds by the paddle trawlers from south of the border, and matters became worse when Aberdeen and Granton followed the English trend. When called to give evidence to the Royal Commission on Trawling they complained bitterly about stock denudation on grounds swept by the steamers. Over the winter of 1884/5 feelings reached fever pitch. A whole series of protest meetings were held in a string of local fishing villages from Nairn to Johnshaven and these were followed in early December by a mass meeting of fishermen in St Katherine's Hall, Aberdeen, before which some 600 or 700 of them had paraded to the skirl of the pipes along Union Street and Castle Street. This meeting condemned trawling and called for its abolition or restriction.[7] It also caused the protest to spread and further mass meetings were held at Peterhead, Fraserburgh and other places. Scottish MPs and the Trawling Commission, which had yet to report, were constantly lobbied on the issue.

The high feelings then prevalent in these strongly Presbyterian communities—already pinched by a collapse of the summer herring trade—were hardly assuaged when they saw these steamers continuing to fish on the Sabbath. Nor did the regular reports of trawlers sweeping away lines help. Not surprisingly, tempers boiled over, most notably at Wick at the end of February 1885. There, bad weather had obliged the local linemen to leave their lines in the sea on a Saturday and on returning to the grounds on the Monday they found that many had been cut or swept away. The steam trawlers seen working on the Sunday were immediately blamed. Later on that day, 2 March, the North Shields steam trawler, *Royal Duke*, sailed into the harbour and moored at the south quay. A crowd gathered and watched the catch being landed but, whilst it was being sold, some of the bystanders took hold of a mooring chain and swept the lot into the harbour. Uproar then broke out and one buyer was struck by a flying stone as his colleagues and the auctioneer took flight. At the same time the steam trawler *Miss Roberts* sailed into the harbour, only to be met by a volley of stones which cut one of her crew on the head. The trawler *Toiler*, also of Aberdeen, had already tied up but as the angry crowd milled around, her mooring ropes were cut. It took the arrival of the Procurator Fiscal and a large contingent of police and coastguards to restore order.[8] Similar feelings about steam trawling also manifested themselves amongst the line fishermen of north-east England.

Before the 1880s the trawling interest, almost to a man, had, of course, stuck to the belief that their mode of fishing was not harmful to stocks. Gradually circumstances forced the smackowners to alter their assessment of the situation. In 1878, for example, the leading Scarborough smackowner, James Sellers, had supported the view that the supply of fish and landings by his boats were as good as ever.[9] By 1885, however, he was expressing his belief that there had been a gradual falling off in the amount of fish his men brought in.[10]

A similar shift was discernible at Hull. In 1878 John Sims, a leading smackowner and president of the local owners' insurance association, spoke for most of his associates when he dismissed the idea that catches were falling off. He stated that the supply of fish was double what it had been twenty or thirty years previously and his views were echoed by the leading smackowner, Alfred Wheatley Ansell.[11] Yet when Ansell went before the 1885 Royal Commission on Trawling he was prepared to admit that catches were falling on traditional grounds and that trawlers were being forced ever further afield.[12] Two years later the same point was expressed more strongly when Mr Ashford, manager of the Hull Trawl Fishermen's Protection Society, publicly criticized the views of Huxley and alleged that the grounds were being over-fished due to the reckless manner in which they were being worked.[13]

Though the 1885 report of the Royal Commission—from which Huxley had to withdraw through ill-health—did concede that trawling on narrow grounds and in inshore waters could affect stocks, its report was still a long way from accepting that man's activities could really damage the industry's entire resource base across the North Sea. The report included the first scientific study, by Professor Mackintosh of St Andrew's University, into the effects of trawling on stocks. Though his investigations strongly influenced the findings of the report they did not assuage the fishermen of his home town who promptly demonstrated against them and reportedly burnt him in effigy.[14] Whilst all previous investigations had been hindered by the lack of statistics, one lasting result of the report's recommendations was that the Government finally agreed to set up a national system for the collection of fishery statistics. Though this has needed considerable refinement and improvement over the years it has at least provided invaluable data for those determining the health of fishery stocks and subsequent fisheries' policy.

The 1885 Royal Commission also recommended that a central body be set up to administer the United Kingdom's fisheries. A survey of the bodies concerned in some way or another with the fisheries at the time reveals a complex and inconsistent picture. Both Scotland and Ireland had fishery boards, though the latter had more substantial powers. The precursor to the Scottish Board had originally overseen certain fishery operations in England as well but had been confined largely to Scotland since 1850. England and

Wales lacked a central authority and various government departments had responsibility for aspects of fishery operations.

In 1885 the powers of the Scottish Fishery Board were widened and extended and the following year the small Salmon Fisheries Inspectorate—which had been largely charged with overseeing the fresh-water fisheries—was transferred to the Board of Trade. Its powers were widened to include the preparation of reports on English and Welsh sea fisheries and the collection of the wider range of fishery statistics. In 1903 it was to move departments once more and, henceforward, the Board of Agriculture and Fisheries took responsibility for superintending the English and Welsh sea fisheries.

Meanwhile, smack profits continued to decline and their owners were soon pressing for wider forms of regulation to aid the conservation of stocks. A fisheries pressure group called the National Sea-Fisheries Protection Association had been formed in 1882 and stock depletion was high on the agenda of most annual meetings. Its 1888 conference, which drew owners from all ports, was held in London and carried a resolution calling for the fisheries authorities to be given power to suspend or regulate trawling and certain other methods of fishing whenever expedient.[15] The trade's leaders held a further conference on the subject the following year and once more complaints were aired about the large and distressing diminution of soles, turbot, plaice and other flat fish in the North Sea. It urged that negotiations be started with Continental countries with the aim of agreeing some international means of regulating the fishing grounds.

Two further conferences were held on the issue in Hull and London in 1890. The former of these took place in April and included delegates from the North Sea trawling ports of Scarborough, Hull, Grimsby, Yarmouth, Lowestoft and Boston. They tried to agree some form of self-regulation to prevent fishing in the summer on grounds which had suffered from 'the wholesale capture and destruction of immature and inedible fish', and a so-called 'self denying ordinance' was passed by which all bodies represented would abstain from fishing on the grounds during the coming summer months. But without the backing of legislation the grounds were never effectively closed and the initiative foundered. The same conference also called on Parliament to impose restrictions on the sale of immature fish and mandated their delegates at the London summer conference to press for 'legislative interference of a national and international character'.[16] In July 1891 an unprecedented international conference was held in London and attended by representatives from Belgium, Denmark, France, Germany, the Netherlands and Spain. This not only called for a convention on the landing and sale of immature fish but also for more scientific and statistical information.[17] The following year saw yet another conference held on the same issue.

Further government legislation, passed in 1888, allowed for the creation of district sea fisheries committees around the coasts of England and Wales and several of these passed by-laws banning trawling within the three-mile limit which was the boundary of their jurisdiction. The Scottish Fishery Board used its new powers to close most territorial waters under its jurisdiction to trawlers. The Board soon went further and in 1889 banned trawling by British vessels in the Moray Firth even outside the three-mile limit. In 1892 the area was extended until about 1,480 square miles of the Firth were closed. However, a number of Grimsby trawlers found a way around this by re-registering in other countries—notably Norway—which were not touched by such legislation.[18] In 1897 a pioneering Danish joint-stock steam trawling company 'Dan' took advantage of the ban on British registered trawlers and sent its steamer *Dania* to work these waters.[19] The venture proved lucrative for several years much to the chagrin of the British.

The difficulties in the way of securing international agreement on regulating North Sea fishing were formidable. The statistical information becoming steadily available to the British Government was by no means sufficiently sophisticated or long-term to provide conclusive indicators on what steps were actually needed. In addition, the concept of international co-operation was still very much in its infancy and at variance with the tradition of the freedom of the high seas. Though several states around the North Sea did take conservationist steps at the time, these were largely unilateral and usually concerned fishing within territorial waters or the banning of the sale of immature fish on home markets.

Given the size and sophistication of the British trawling trade, and the potential pressure that could have been brought to bear, it seems possible there could have been considerable movement towards international regulation and conservation during the 1890s had the will to carry it through been maintained. This would have been to the long-term benefit of all. However, technological developments helped the British trawling trade find another way out of the crisis and diminished its interest in international regulation.

Traditionally, the trade had viewed trawling primarily as an extractive activity and had sought to maximize output not only by markedly expanding the size of the North Sea fleet but also by continuing to enlarge and improve the catching efficiency and range of each individual smack, as well as by applying more intensive methods of fishing, including fleeting. By the 1880s it was evident that sailing smack design was nearing the limits of technological development. The minor improvements that could be made were insufficient to compensate for the decline in stocks and the new grounds which were being opened up were at the very edge of a smack's working range. If the trade was confined by the limitations of the sailing trawler then it would have been imperative that it sought a means of conserving the depleted North Sea stocks. However, the 1880s also saw the development of the purpose-built steam-screw trawler. Though initially confined largely to the North Sea they were soon to prove their worth. Not only were they able to haul larger gear but they were less subject to the vagaries of wind and weather than the sailing smack. By the latter half of the 1880s it was already evident that they were far more efficient fish-catching machines than the conventional smack. Had they remained tied to the North Sea then they could at best have merely delayed the need to address the problem of stock denudation for a few more years; it was a cut in catching effort not an increase in catching efficiency that was really needed. But steam trawlers were soon being built with a far greater working range than the sailing smack. This would allow the leading British trawling ports to break out of their dependency on near- and middle-water grounds and enable them to exploit fish stocks on new grounds in distant waters. It would also allow them to lessen their clamour for conservationist policies.

During the 1880s sailing smacks had worked as far north as Shetland and the depletion of North Sea stocks pushed some steam trawlers further afield in pursuit of new grounds. In the early 1890s vessels from Hull and Grimsby were trying their luck off the west coast of Ireland and in 1893 a Hull fleet was reported to be fishing for hake in the Bay of Biscay and running catches into Plymouth.[20] Later, trawlers were to venture down to the African coast. However, those vessels which voyaged in a northerly direction, following in the wake of the line fishermen, proved infinitely more successful. British lining vessels had, of course, visited Faroe and Iceland for centuries and there was an increased activity in that direction from Grimsby in the 1880s. The discovery of the splendid halibut lining grounds known as the Faroe Banks was followed some years later by a renewed interest in the southern coasts of Iceland by steam liners.[21] For some years such grounds were to be the major source of summer halibut.

Trawling around Orkney and Shetland had at first yielded only poor results (though Scottish vessels were more immediately successful than their English counterparts) but eventually grounds east of Shetland proved more rewarding. Some English steam liners may have experimented with trawling off Iceland as early as the summer of 1889. The first steam trawler to try its hand there was the German vessel *President Herwig* but the voyage was dogged by ill-luck for the crew seem to have trawled over a submarine lava field which tore their trawl to pieces and forced them to return home without a catch.[22]

In the summer of 1891 the steam trawler *Aquarius*, owned by the Grimsby Steam Trawling Company and skippered by Mr T. Cotton, sailed for Iceland and trawled off Ingol's Hofde Huk (Ingol's Hook according to Humber pronunciation) and returned with a good catch of plaice and haddock. The following summer some nine trawlers went to the grounds and filled between 100 and 400 boxes per trip, principally of plaice and haddock but also some red fish and halibut.[23] These steam trawlers were working at the very limit of their operational range and coal was stored not only in the bunker and fish-room on the outward journey but also in every available place. Larger vessels, designed primarily for the rigours of the trip, were soon being built but at first Iceland remained primarily a summer fishery as the weather, particularly the fogs, made winter voyages hazardous. Initially, most trawlers limited themselves to working off south-east Iceland or at any rate no further from home than the Westmann Islands. In winter many vessels concentrated on Faroese grounds but, after pressure, the marine insurance companies gradually eased their restrictions on winter voyages[24] and vessels from several nations were soon working right round the island. By 1903 between sixty and seventy Grimsby trawlers were visiting Iceland on a regular basis, together with about fifty-five steam lining vessels, and a further eighty Hull trawlers were engaged in the trade.[25]

From the start, this attention from foreign trawlers caused great alarm in Iceland, especially as some of the visitors seemed to have scant regard for the three-mile territorial limit. Iceland, like the Faroes, was part of the Kingdom of Denmark and the islanders called for greater vigilance from the Danish Navy in policing these waters. Iceland was then a poor country and the fish stocks were its most important natural resource. The coastline around the Bay of Faxa was the most heavily populated part of the country but many of the grounds worked by local fishermen were outside territorial limits and easy prey for foreign trawlers. Like many linemen before them they feared for the viability of such grounds when heavily worked by trawlers, and the denudation of North Sea stocks which had originally encouraged the voyages to Iceland only served to increase their worries.[26] Even before trawlers began to arrive steps were taken to counter their effects. In 1889 the Danish Government submitted a bill to the Althing prohibiting trawl fishing within territorial waters.

The British deep-water trawling trade, recently so keen on encouraging state regulation of fishing activity in the North Sea, began to change tack as the Icelandic grounds grew in importance. Some sections of the trade had previously supported calls for territorial limits to be extended and trawling banned in certain areas of the North Sea in the interests of conserving stocks. By the later 1890s, an Icelandic plea for their exclusion from the rich grounds in the Bay of Faxa was viewed in a completely different light by many British trawler owners. The best that some of them were likely to consider was a reciprocal agreement in which British trawlers would stay out of the Bay of Faxa in return for being allowed within the territorial limits on another part of the coast.[27] However, as the British trawling trade became more and more reliant upon distant-water grounds close to the shores of other nations it realized that any moves to alter territorial limits could be detrimental to its newly developed interests. This view was perhaps best expressed later by Charles Hellyer, the leading Hull Trawler owner, to a government commission in 1909:

> Seeing that the number of British fishermen is at least six to one of any other nation, it is of paramount importance that the three mile limit be maintained, and that Great Britain do nothing whatever in making representations to any other nation whereby their fishing vessels might be influenced not to fish in waters which are international and so give a viable pretext for other nations to close, or prohibit British fishermen from fishing in similar waters off their coasts. I do not think I can put it much clearer than that. We feel very strongly about that, we feel very strongly that any authority . . . trying to make terms or to enter into negotiations on any matter affecting the three mile limit, will be disasterous to the fishing and mercantile interests of this country . . . We say further that any such tampering with the three mile limit is in our opinion pregnant with international difficulties . . . because we have to approach other people's shores to bring the fish to England.[28]

By that date the leading sector of the trawling interest had invested heavily in the practice of distant-water fishing and increasingly relied upon the exploitation of fish stocks on grounds beyond the North Sea: grounds close to foreign coasts. The trawling interest would only support North Sea conservation proposals if they did not question existing concepts of territorial waters and thus fishery limits.

Broadly speaking, fishery limits—the areas of sea over which a state had an exclusive right to make and enforce regulations for catching fish—and territorial waters were then the same, in terms of the areas covered. Yet until the 1830s there had been little international agreement about what were the exact limits of territorial waters. Many states bordering the North Sea had traditionally regarded three miles—supposedly the range of a cannon shot from shore—as the normal limit of a state's jurisdiction. However, it had never been clear whether the three miles were measured from the cliff top, high- or low-water marks and whether bays and estuaries ought to be included. An 1818 Convention between Britain and the United States and another between Britain and France in 1839 attempted to clarify the issue and defined territorial waters—and fishery limits—as covering the sea area stretching three miles from the low-water mark. In the case of bays less than ten miles in width then the boundary was to be a straight line reckoned in a similar manner from the low-water mark of the two headlands. This definition for North Sea waters had also been incorporated in the 1882 Hague Convention on the policing of fisheries in international waters. Thanks in no small measure to Britain's position as the world's leading naval power and the skill of her diplomats, her concept of international waters gained wide acceptance by the early twentieth century.

The 1889 Act prohibiting trawling in Icelandic territorial waters actually made no attempt to define the term 'territorial waters'. Traditionally, the Danes had maintained a four-mile limit but had been party to the 1882 Hague Convention which set North Sea territorial limits at three miles. In the 1890s the situation regarding Icelandic limits was less clear and Danish patrols tended to oversee waters which would fall within a three-mile limit.[29] The growing British trawling presence off the Faroese and Icelandic coasts nurtured native resentment. They were accused of taking little notice of Icelandic fishermen, of damaging their gear and driving them from the grounds.[30]

The 1889 Act was proving of little use in protecting Icelandic territorial waters, and trawlers continued to 'explore' bays and fjords, often working close to the shore. In 1894 Iceland passed a much stronger Act. This sought, in essence, to prevent vessels with trawls on board even entering territorial waters or putting in at Icelandic ports except in cases of emergency. Strong penalties were laid down including the confiscation of gear. This Act represented a remarkable step as it was the first time that the Icelandic Althing had taken the lead in a matter concerning relations with other countries without first consulting the Danish Government.[31]

The British trawling trade protested strongly against this Act. They argued that it was not only harsh but also contrary to the spirit of existing international law. Over the next few years a string of trawlers—not all British—were arrested, fined and had their gear confiscated for illegal fishing off Iceland, and also the Faroes. Complaints from the trawler owners continually flowed into Westminster. The issue soured relations between Britain and the mother country Denmark. In 1896 the British Government sent a Royal Navy Training Squadron on a visit to Iceland under the command of Commodore G. L. Atkinson. Atkinson's squadron repeated the visit in 1897 and he developed a good working relationship with the Icelandic Governor-General Magnus Stephensen. The accord they struck established a basis for discussions which led to the Anglo-Danish Territorial Waters Treaty of 1901.[32] Under its terms the Danish Government accepted the three-mile definition of territorial waters around Iceland, the Faroes and Greenland. Though relations showed some improvement thereafter and trawlers were once more able to visit Icelandic ports, the Treaty was a source of long-term resentment for Icelanders who felt that the Danes lacked a real concern for their fishing industry and had sacrificed their interests in order to allow Danish dairy products easier access to the British market. Indeed, the definition of territorial limits which they were forced to accept left many of Iceland's broader bays in international waters. Such resentment later encouraged drives for independence.

Though the North Sea remained a major source of white fish, it declined in importance, especially after the turn of the century. The figures in Tables 1 and 2 illustrate that though the North Sea trawling effort—in terms of what Garstang defined as 'smack units', that is basically the number of smacks or sailing trawlers or their steam equivalents—increased by about 250 per cent during the 1890s, the catch by each fell by nearly a half. Garstang's research on the 1890s period indicates that the North Sea stocks fell by a half over the decade and that the reduction in stocks was due to the increase in the level of trawling effort deployed.[33] The introduction of the more efficient steam trawlers meant landings continued to increase over the decade but only at the cost of further reductions in stock levels. As stock levels decreased fewer large fish were caught and landings comprised an ever greater proportion of immature fish which had not reached their breeding potential and were thus unable to add to the stock. Catching more and more of an ever-diminishing stock had obvious results and after the turn of the century an absolute decline set in. United Kingdom white fish landings continued to grow rapidly: the 330,000 tons landed in 1903 had grown to 420,000 tons by 1913 but the North Sea's contribution of demersal fish fell from 4,351,429 cwt in 1905 to

3,551,756 cwt in 1913. The demand from fish friers was growing rapidly at this time and distant-water fish helped satisfy this demand.[34]

Table 1. Annual Average Catch of Four Grimsby Trawling Smacks 1867 and 1875–1892 (in cwt)

Year	*Plaice*	*Haddock*	*Prime*	*Rough*	*Total*
1867	998	831	63	46	2012
1875	549	937	63	30	1565
1876	601	891	50	33	1576
1877	421	668	88	21	1198
1878	254	481	76	31	843
1879	298	488	98	44	928
1880	291	359	65	39	754
1881	242	280	84	70	765
1882	385	717	84	86	1273
1883	340	665	97	74	1177
1884	325	526	96	79	1025
1885	280	477	90	89	936
1886	250	510	77	87	925
1887	221	475	62	87	846
1888	195	372	42	57	667
1889	177	342	64	69	652
1890	205	465	47	65	783
1891	203	590	47	79	920
1892	168	436	29	49	683

A quinnquennial summary of the preceding figures

1875–9	425	693	75	32	1222
1880–4	317	509	85	70	981
1885–9	225	435	67	78	805
1890–2	192	497	41	64	795

Table 2. Catches of all Bottom Fish, Catches per Smack Unit (in tons) and the Number of Smack Units 1889–1898

Year	*Catch*	*Smack Units*	*Catch per Smack Unit*
1889	173,810	2859	60.6
1890	172,055	3086	55.7
1891	180,054	3711	48.5
1892	187,512	4057	46.2
1893	200,281	4307	46.5
1894	215,408	4599	46.7
1895	228,180	4918	46.4
1896	232,034	5620	41.3
1897	225,864	6099	37.0
1898	230,656	7143	32.3

Garstang's definition of 'smack units': Garstang based his unit on a sailing smack. The efficiency of steam trawlers and their use of the otter trawl meant they caught more fish. Garstang estimated four times as much fish as a smack of the same weight and built this difference into his figures.

Source of Tables 1 and 2: W. Garstang, *The Impoverishment of the Sea*, 1900. Table 2 also quoted by Cushing, D., *The Arctic Cod, 1966.*

The Icelandic grounds were now being fished throughout the year and in 1905 the Barents Sea was visited for the first time. Though both fishing grounds necessitated five or more days of steaming time, the catch, once there, was much higher than in the North Sea. In 1906, for example, the catch per day from Iceland and the Barents Sea was 44.2 cwt and 40.2 cwt respectively, whilst that from the North Sea was only 17.6 cwt.[35] By the First World War trawlers had made trips to the Grand Banks, Newfoundland and the Greenland coasts. The steam trawler made possible the large-scale exploitation of distant waters, yet its introduction and subsequent development necessitated a radical restructuring of the trawling trade.

From the later 1880s sailing smacks gave way to steamers at an ever-increasing rate. The last new smacks were built for Hull and Scarborough in 1886 and Grimsby in 1893. Though they continued to scratch a living for a time on the North Sea grounds they became less and less profitable. The introduction of the otter trawl in 1895 speeded their departure from northern ports. The otter trawl did not require the unwieldy beam as the net mouth was kept open by so-called otter boards. Initially, these were attached to both wings of the net mouth and connected by thick wires—warps—to a winch barrel on deck. The gear had been worked experimentally from yachts in the 1880s but its first commercial application had been pioneered by a Mr Scott of Granton in 1895. Once proven successful it was adopted by steam trawlers up and down the coast and on parts of the Continent within the year. It was estimated to be some thirty per cent more efficient than the beam trawl but proved difficult to adapt to sailing smacks which thus failed to benefit.

Steam trawlers were altogether more efficient catchers of fish. It has been estimated that they are likely to take four times as much fish in a year as a smack of the same tonnage. The remaining smack fleets along the north-east coast soon disappeared. Those lost in the Great Gale of January 1895, which took the lives of 106 Hull fishermen, were not replaced. In 1884 Hull had possessed more than 450 smacks and yet, like Scarborough, it had disposed of virtually every one by 1900. The last Grimsby smack sailed in 1903. The second-hand market was so flooded that they proved difficult to sell. Many were broken up, a few were converted for coastal trading whilst others were sold abroad. The Scarborough iron smack *Contrast* found its way to Dakar in French West Africa.[36] Many others were bought cheaply by the Faroese, Norwegians and Swedes. Though displaced in England they proved technically superior to most of the traditional sailing boats in their new home ports. Many were used in the Faroese hand-line fishery and had a considerable impact on traditional patterns of activity in the islands.[37] Many of the smacks which found their way into Scandinavian ownership were later fitted with diesel engines and deckhouses and some worked through to the 1970s.

Not all ports followed this trend. Though Aberdeen, Fleetwood, Grimsby, Hull and Shields trawled almost exclusively by steam, a few centres retained smacks; but those that did were no longer in

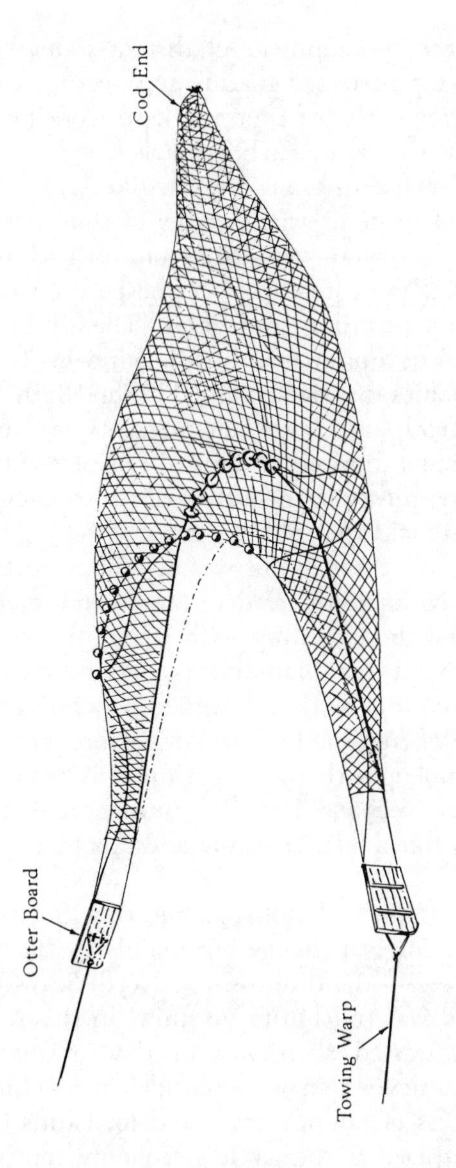

Fig. 3. The Otter Trawl

the mainstream of the trade. A few were still being built for Brixham and Lowestoft—ports far from the coalfields—as late as the 1920s.

Steam trawling remained a capital intensive affair. Even in the early days, a basic second-hand steam paddler had cost twice as much as the biggest new smack off the stocks and costs continued to rise with the development of distant-water vessels. Moreover, they required sophisticated marine engineering back-up to remain operational. Previously, of course, a hard-working fisherman could eventually raise the capital to buy a smack and in the days of sailing trawlers the fleets sailing out of ports like Hull and Grimsby were in the hands of quite a large number of small owners. In the new distant-water fleet there was little room for such owners and many went to the wall whilst others became partners or shareholders in the new steam trawling companies. The distant-water trade in particular became dominated by the kind of joint stock limited liability companies that had emerged in the 1880s. These companies were empowered to raise large amounts of capital and often recruited cash and directors from other areas of business and commerce. The highest position most distant-water fishermen could expect to aspire to in the future was that of skipper.

The shift from steam to sail was accompanied by a fundamental restructuring of ancillary trades. Metal and marine engineering trades displaced those dealing with wood and wind. There was an increased demand for blacksmiths, engineers and riveters and correspondingly less work for ship's carpenters, sail- and block-makers. Many of the smaller boatyards which had built the smacks could not compete with the iron shipyards of Aberdeen, or those on the Humber, Tees and Tyne. In a number of respects the trawling trade in ports like Hull, Grimsby and Aberdeen typified the new industrial age.

The forces which had reshaped the trawling trade were driven partly by technological change but mainly by the need to overcome resource problems—the diminishing North Sea white fish stocks—whilst demand was continuing to grow. In this respect, the trade's experience contrasted sharply with that of many other British economic activities which were obliged to re-adjust in the face of a seeming excess of raw materials and foodstuffs flooding into the British marketplace as the world's economy continued to grow. It was not until the 1890s that the exploitation of the middle- and distant-water grounds with steam trawlers gave the trade a strategy for breaking out of its reliance on diminishing British coastal and North Sea fisheries. But this was a strategy based upon extraction and exploitation of fish stocks rather than one of management and conservation.

9

Steam and Storm

For the trawlermen some of the most radical changes of the new steam and distant-water era were related to payment, crewing and handling of craft. A typical late nineteenth-century steam trawler carried a crew of ten whilst a smack usually shipped five. Their steam engines could only be operated by personnel with technical skills and each vessel generally carried a chief engineer, second and fireman trimmer. On the Grimsby and Hull smacks only the skipper and mate had definitely been 'on shares'. Before the early 1880s at least two of the remaining three crew members had usually been unwaged apprentices but, when the pool of indentured labout began to dry up, their place had been taken ny weekly-paid hands.[1] Part of this tradition was carried forward onto steam trawlers. In 1900 the usual Humber steam trawling practice was to pay just the skipper and mate by results. Under the Grimsby system, for example, they took 1 3/8 and 1 1/8 out of a total of fourteen shares respectively. The rest of the crew were provided with free food and a basic wage which ranged from forty-six shillings for a chief engineer down to slightly over £1 for a deckie.

Having such a large proportion of the crew on wages was quite unusual in many other sectors of the fishing industry where the uncertain rewards had been traditionally determined by the share. In June 1901 the Grimsby owners, facing increased prices for coal, trawl nets and other gear, suddenly announced they were introducing an element of the share system into the pay structure of all crew members. This arbitrary action by the owners provoked an equally spontaneous reaction from the Grimsby trawler engineers. As they came in from sea they decided not to sign on again and a bitter dispute began which the owners regarded as a strike and the men as a lockout. Few trawlermen were working and the port's fish trade ground to a virtual standstill over the summer. By September many fishing families were suffering considerable distress and feelings were running high. When a number of owners under Henry Smethurst attempted to man their vessels with foreign crews the waterfront erupted. A mob stormed, wrecked and then fired the offices of the new Trawler Owners' Federation whilst the officials fled on to the roof and only escaped through a neighbouring building. Rioting spread swiftly out from the fish dock and along the Cleethorpes Road where many properties were damaged. As police reinforcements arrived in the town they were stoned and when they replied with a baton charge several bystanders were injured. The Riot Act was read by Alderman Hewson in Riby Square and order was roughly restored by the 300 policemen drafted in from Manchester and Sheffield. A battalion of infantry were held in readiness in the barracks nearby.[2]

The dispute soon highlighted the differences that had developed amongst fishermen in the steam-trawling era. The engineers were well organized in the mould of skilled-engineering trade unionists. They paid high subscriptions, collected good benefits and were usually controlled by a moderate leadership that had been taken somewhat off guard by the spontaneous action of their membership at the opening of the dispute. In contrast, the deckhands had no comparable degree of organization. The zenith of nineteenth-century trade unionism amongst fishermen had been reached in 1890 when Reuben Manton of Hull had helped lead moves to combine various fishing unions and associations with those that had been set up—generally by skippers and mates—at other ports. The National Federation of Fishermen of Great Britain and Northern Ireland was formed at a time when there was a great surge of unskilled trade unionism across Britain. It soon claimed to have 5,000 members.[3] In the later 1890s, like many similar associations, it experienced a rapid decline and Reuben Manton later became principal of the first Nautical College in Hull. The union found it particularly difficult to maintain the deckhand's interest in its ideals and still harder to collect subscriptions amongst crews landing on many tides in a host of different craft. The vagaries, uncertainty and sheer drudgery tended to concentrate many trawlermen's minds on the pleasures of the here and now when ashore and not on organizing for the future. Those who did have an inclination for leadership often became skippers and mates and duly formed their own protective association. The unique problems associated with organizing trawlermen were to beset the trade union movement for generations to come.

Just before the Grimsby dispute had broken out an attempt had been made to fill the void left at Grimsby by the decline of the National Federation of Fishermen. The Gasworkers Union had managed to recruit about ten per cent of the deck crews there and sent Walter Wood across from Leeds to organize them. The National Sailor's and Firemen's Union, led by Joseph Havelock Wilson, also claimed members in the port. In the 1880s and early 1890s Havelock Wilson had pursued a militant approach but by this time was a marked moderate and was soon calling for the men to return to sea.[4]

In contrast to the various labour organizations, the Federation gave the Grimsby owners an apparently united front—though some employers did not join them and continued to fish throughout the strike. Though the owners had suddenly announced the introduction of the share element they had not acted impulsively. July and August were known to be months when the market was so flooded with fish that the owners would lose little by laying their trawlers up in dock. They let it be known that even if the dispute ended swiftly they were unlikely to send all vessels straight back out to sea and turned down requests to open their books for inspection. All calls for arbitration were treated with suspicion for, in their eyes, external intervention in labour matters usually proved detrimental to their perceived interests.

The violence at Grimsby attracted nationwide attention but the dispute was not resolved until October when Sir Edward Fry was finally brought in to arbitrate. Initially, his award seemed more favourable to the men for although it included an element of the share system it gave them a guaranteed minimum wage and supported the men's contention that they should sign on at the Board of Trade offices rather than those of the Trawler Federation.

Many trawlermen and outside observers saw the award as a defeat for the owners and that the guarantees it embodied meant that wages would fall little if at all. In retrospect, however, it can be seen that the owners gained a vital concession and had ensured that the share element would determine part of a trawlerman's earnings. Once this was accepted it stayed and the system gradually spread to all other steam-trawling ports and, over time, determined a greater portion of a deep-sea trawlerman's wages, whether skipper, mate, chief engineer or deckhand.[5] The reorganization of the trade's labour resources, in the wake of the shift to steam and more distant waters, was now well underway.

The replacement of sail by steam may have reduced the number of fishermen who lost their lives but the difference was at first only marginal. There were still plenty of smacks around and though fewer steamers were wrecked, when they did go down they took a larger crew with them. Trawling remained the most dangerous of occupations and the 1890s were marked by several bad North Sea storms which took a constant toll of smacks and steam trawlers alike. The worst was probably the one which blew up on 22 December 1894 catching large fleets out on the grounds. The subsequent loss of life was tremendous, probably worse than in the great storm of 1883. Hull alone lost six steam trawlers and fish cutters as well as eight smacks. Tide after tide, the St Andrew's Dock lock-gates were besieged by crowds of anxious relatives waiting for some news of the missing vessels. All told, some 108 Hull trawlermen perished in this storm, leaving behind them 55 widows, 106 children and 15 aged dependants.[6]

Few North Sea trawling ports were unaffected and Hull's Mayor suggested to the Lord Mayor of London that the Mansion House might open a fund for the relief of the destitute widows and orphans. The Lord Mayor of London replied that, while he sympathized deeply with the sufferers, the Mansion House funds were 'generally associated with exceptional national calamities and it was always difficult to appeal to the national pulse and purse in matters which constantly recurred'.[7] Local appeals, he deemed, were the best and these were soon started, but the one in Hull raised no more than £2,500[8] for the numerous dependants—or considerably less than half the cost of one steam trawler.

In some ways the Victorians saw shipwrecks as their descendants were to view car accidents. Though they sometimes inspired dramatic hymns, paintings and stories in a way in which road accidents never could, the relentless frequency of the sea's carnage reduced the scale of its impact. Other seafarers often regarded trawlers as the 'lifeboats of the sea' for the trawlermen's rough and dangerous life made them fearless and skilful rescuers. On countless occasions they risked life and limb to pluck other seafarers from a watery grave. In January 1895, for example, the German liner *Elbe*, en route from Bremen to New York was struck amidships by the small Aberdeen steamer *Crathie*. It was 5.30am on a bitterly cold and dark morning with a heavy sea running and the *Crathie*, her bows stove in, lost touch and limped on to Rotterdam; the *Elbe*, meanwhile, had been mortally damaged. The liner began to settle quickly in the water as half-clad and panic-stricken passengers rushed onto the decks.

Confusion reigned, attempts were made to launch three lifeboats in the heavy seas but only one seems to have got away. This lifeboat dragged in the only woman survivor, Anna Boecker, who had been thrown into the ice-cold sea when one of the other boats capsized. The tiny craft drifted throughout the day as the crew fought to keep its head to the wind. All twenty survivors suffered dreadfully in the biting cold. Waves continually broke over the boat, and their scanty clothing was soon covered with ice. Several vessels passed by on the horizon but failed to notice their plight and things looked desperate until they were spotted by the Lowestoft smack *Wildflower*. Rough seas made it difficult to close but *Wildflower*'s skipper, Billy Wright, skilfully brought the smack near enough to throw lines across. After several attempts one was secured and the lifeboat was pulled to the smack's lee side. About half the occupants were able to jump on board before the line parted and the lifeboat drifted away once more. Wright stuck to his task, bringing the *Wildflower* round again and another line was made fast. Most of the rest were then dragged on board but Anna Boecker was lying on the bottom of the lifeboat in a bad way. Wright and another man scrambled down and helped her into the smack before the line parted for good and the lifeboat drifted away.[9]

The grateful survivors were taken below and given warm food, drink and clothing as the *Wildflower* ran for Lowestoft with news of the disaster. A search was soon undertaken but no more survivors were found and the remaining 375 passengers and crew were given up as lost. Occasionally fishermen brought up bodies in their trawls and, whilst never becoming hardened to this experience, usually returned them to the sea; on this occasion they were paid to bring them ashore and for several weeks bodies and wreckage were brought in until the advanced state of the corpses' decomposition rendered this gory job totally unpalatable. Sometime after the disaster a smack was making for Lowestoft as night was falling when those on watch spotted some mysterious figures dancing up and down on the water. Closer investigation revealed them to be six lifebelted corpses in perpendicular position and this ghastly sight so terrified the crew that they crowded on sail and put as much distance between themselves and the spot as possible. They were still in a state of deep shock when they tied up at Lowestoft.[10]

Though trawlermen had saved many lives in similar circumstances, the *Wildflower*'s rescue caught the public's imagination. Wright became a national hero overnight. A grateful Kaiser presented him with a clock and his four crew with silver watches whilst a subscription raised by a London journalist enabled him to buy his own smack which he named *Willing Boys*.

When trawlers got into difficulties far out on the grounds they could often only expect help from their own kind. On 25 January 1910 the Hull steam trawler *Gothic* was out on the North Sea dodging through a blinding blizzard and mountainous, roaring seas whipped up by hurricane force winds. About 4am a huge wall of sea swept the trawler from fore to aft, smashing the bridge and carrying away the funnel, open boat and masts. The crew, who had clung on to anything they could find, were left amongst a mass of wreckage with a sinking ship and a fight for their lives. They had to pump by hand after the steam pumps failed as the engine room flooded. Everything available was soaked in oil and burnt as distress flares. The ferocity of the blizzard kept the sky dark throughout the day and, despite a frenzy of pumping and bailing, the *Gothic* continued to sink lower and lower in the water.

That afternoon the Grimsby steam trawler *Oldham* saw their plight and came up. The storm had not eased but directly his vessel was close enough, the *Oldham*'s skipper, Sidney Marshall, launched the open boat and, after a terrible struggle, reached the stricken trawler. Four men jumped on board before they pulled back to the *Oldham*. They had scarcely scrambled to safety when a huge wave smashed their little boat to pieces against the trawler's side. For hours Skipper Marshall tried to get as close as he could to the *Gothic*

Plate 3. Icing up in the Barents Sea

and eventually got a line across. Lifelines were then passed over and the crew lashed them round their bodies—all that is except the cook who had collapsed and had to be lashed in by the mate. By now the *Gothic* was well down and seemed likely to go under at any time so the crew jumped overboard and were dragged through the raging seas and on to the deck of the *Oldham*. All survived this horrific ordeal except the cook who was found to be dead when pulled on board. Within half an hour the *Gothic* finally slipped under, some twenty hours after being hit by the wave.[11]

Steam trawlers making trips to Iceland may have expected less help if they got into difficulties off its rugged, remote and seemingly inhospitable shores, but if they did so then they underestimated the generosity of the Icelanders. In the early years, when only summer trips were undertaken, few trawlers seemed to have been wrecked off Iceland but once they started fishing there throughout the year it was a different story; between 1899 and 1914 at least twenty-eight Hull steam trawlers alone were lost on Icelandic trips. On several occasions crews found themselves cast ashore apparently miles from anywhere.

At 11 pm on 13 January 1910, for example, the steam trawler

Thomas Hamling, owned by the Neptune Steam Fishing Company of Hull, was stranded off a low shore some 30 miles east from the Portland lighthouse. The sea was fierce and waves kept breaking over the stricken vessel throughout the night. In the morning two men, the mate and the cook, tried to get ashore with lines but only the mate, Mr Copeman, succeeded. The cook, J. H. Metcalfe, was overwhelmed by the waves and drowned in full view of the helpless crew. Once the mate had secured his line, the rest of the crew scrambled along it to the shore with the skipper being last to leave. Without shelter or provisions, they were in a sorry state and one of the deckhands, Robert Gibson, had a broken leg. Metcalfe was laid in a temporary grave and a makeshift stretcher was constructed out of an old rug and two poles. The survivors then took turns to carry Gibson as they walked and scrambled all day through the wild and inhospitable country. By evening they were utterly exhausted and a decision was made to press on for help without the injured man who, despite his entreaties, was left as well-sheltered as was possible behind a hillock. About half an hour later they stumbled across an Icelandic farmhouse and as soon as the farmer could summon further help he and the skipper went back for the injured man, but by then another Icelander had come across Gibson and taken him back to his house where a doctor was called to set his smashed leg.

The crew found the Icelanders poor but extremely hospitable. Their farmhouse, constructed from wood and stone, was very basic but warm and the farmer and his family made them extremely welcome and shared out winter rations and spare clothing they could ill-afford. They also gave Metcalfe a proper Christian burial. Most of the crew took a week to recover their strength but Gibson had to spend a further ten weeks there before he was fit to travel. The main party were supplied by the Icelanders with twenty-five ponies, and guides who took them on a wondrous eleven-day, 200-mile journey across snow and ice, by hot springs, through streams and over mountains, to Reykjavik. Wherever they stopped for the night they were treated with overwhelming generosity and there seems to have been little sign of the hostility with which the Icelanders had originally greeted the arrival of trawlers off their shores. At Reykjavik the *Thomas Hamling*'s crew bade goodbye to their guides before catching the Mail Steamer for Leith.[12]

In December of the same year another Neptune Steam Fishing Company trawler, this time the *Mackenzie*, was wrecked in almost the same spot. One of the *Thomas Hamling*'s crew was on board and he faced a repeat of his previous journey across Iceland and thence home.[13]

Though the North Sea grounds gradually declined in importance as steam trawlers increased their distant-water trips, they were to be the scene of the most dramatic event to afflict the trade during the Edwardian era. This was the so-called Dogger Bank incident in which the Russian Grand Fleet shelled Hull's Gamecock fleet of trawlers whilst they worked some 220 miles east by north of Spurn Point. Though the loss of life was minimal in comparison with previous disasters it seemed for a moment that the sheer international outrage provoked by this action might plunge European nations into a major conflagration.

This apparently irrational and totally unprovoked attack on the unarmed trawlermen of the world's greatest maritime nation—in its own backyard—had its origins in an arena of international conflict far removed from the North Sea. In the Far East, Japan and Russia faced a weak Chinese Government and both had designs on the rich resources of Korea and Manchuria. In the Sino-Japanese War of 1895 Japan had won treaty rights in Southern Manchuria, the Liao Tung Peninsula and Port Arthur, which was the most northerly ice-free port on mainland Asia. Yet by 1897 a jealous Russia had secured sufficient international backing to force them out and take over. Whilst Russia celebrated her gain, Japan smarted and thereafter the rivalry between these two expansionist powers intensified and relations deteriorated. Finally, on 6 February 1904, within hours of her diplomats breaking off relations in St Petersburg, the Japanese navy, under Admiral Togo, unleashed a surprise attack on Port Arthur, sinking or damaging many Russian warships and reversing the balance of Far Eastern naval power as decisively as Yamamoto was to do at Pearl Harbour in 1941.

Over the next few months Imperial Russia suffered continual setbacks from a foe she both despised and under-rated. Japan's navy always had the upperhand but, having more limited resources, could not swiftly replace lost ships. It seemed possible that Russia might regain the naval initiative by sending out her Baltic fleet, including four powerful battleships of the latest type, though this would involve a voyage of 18,000 miles and all the problems associated with coaling and repairing a fleet at sea. So, after tortuous preparations, this iron-clad armada of forty-two ancient and modern warships set out on its first leg through the Baltic in the middle of October 1904.

By this time the Japanese had gained such a psychological ascendancy that many Russians seemed mesmerized by their success and expected anything. A belief grew that the Japanese were willing and able to sabotage the fleet well before it could close on Far Eastern waters. There was a widespread belief in Russia that Japan had an espionage network along the shores of northern Europe. As the war had started with the surprise attack on Port Arthur the Russians expected more of the same, even in waters almost 18,000 miles from Japan. Russian counter-espionage sources were charged with uncovering the plan and, anxious to prove their worth, their reports to St Petersburg suggested that the Japanese had built up a suicide detachment, probably made up of torpedo boats and possibly disguised as trawlers, which was lying in wait for the Russian fleet in isolated Norwegian fiords. It was also rumoured that armed merchantmen, submarines and even destroyers were being hastily fitted out in British yards.[14]

The idea of being able to assemble and hide such a force now seems ridiculous, and indeed it is now thought that the Japanese navy did not really possess such a capability and that no spies or vessels had been sent to the area. However, in the rarified atmosphere of suspicion then prevalent in international politics such possibilities did not seem so far-fetched as they now appear. Rivalry amongst European states, especially France, Germany, Russia and the United Kingdom, only encouraged mutual distrust and talk of espionage, conspiracy and covert naval activities along the North Sea's long and often remote coastline. Only the previous year Erskine Childers had based his spy thriller, *The Riddle of the Sands*, on a fictional German plan to invade Britain from the remote low-lying shores of Friesland.[15] For many, including Childers himself, that plot seemed all too plausible and his novel proved almost as popular in its day as a Le Carré thriller would be at the height of the Cold War.

By the time the Russians sailed, the European press and even a few states were taking an interest in these rumours. To the Russians they were now less rumour than reality. Admiral Rozhestevensky continued to receive reports of covert Japanese activity in the North Sea as his fleet edged out of the Baltic. Even two Danish fishermen sent out with a message by the local Russian Consul were driven off when they approached. When they finally delivered the message it proved to be from the Tsar and promoted Rozhestevensky to Vice-Admiral. Tension remained high, lookouts claimed to have spotted reconnaissance balloons whilst intelligence reports talked of unidentified torpedo boats leaving Norwegian waters. By now the Russians had thoroughly convinced themselves that something would happen and soon. The fleet was living on its nerves and losing touch with reality. On the night of 21 October, as cruisers were deployed far to the front of the fleet searching for the enemy, a panicking supply ship reported being attacked by torpedo boats. To Rozhestevensky, on the bridge of the *Suvoroff*, it seemed that the expected enemy warships were closing. In fact the Russian navy was closing on Hull's Gamecock fleet of trawlers.

The Gamecock Fleet had only arrived on the grounds that evening to trawl, and its fishing signals had already warned off several merchant vessels when, shortly after midnight, several large ships were seen steering in loose line-ahead formation. Though they were obviously warships the trawlermen thought they were Admiral Lord Charles Beresford's Channel fleet returning south after visiting Tynemouth. The Gamecock Fleet lay straight in their path and its trawler admiral became anxious as the warships continued to ignore their identification signals and even the green flares they sent up. Anxious to avoid a collision, he hurriedly ordered the fleet to the windward but suddenly the sea was lit up by powerful searchlights and shots started hitting the water in front of the line of battleships. Even then the trawlermen were more concerned to avoid a collision and thought the shots were no more than night gunnery practice. The unthinkable only dawned when the searchlights concentrated on a group of three trawlers which were then enveloped in shell fire. Shocked fishermen ran along the decks waving and shouting; some even held up fish to show what they were.[16]

On sighting the green flares the bridge of the *Suvoroff* had swung its searchlights and guns into action. The emergency 'engage enemy' signal was flashed to the fleet which responded with confused and uncontrolled fire before the searchlights located the 'enemy'. Guns on every ship then swung onto the targets. The steam trawler *Crane* took the brunt of the action. Skipper George Smith and third hand John Leggott were decapitated as shells rained into the ship. The engineer took a serious shrapnel wound to the chest whilst another crew member's hand was blown away. Decks and superstructure were covered with blood whilst the boat winch was smashed to pieces. The *Crane* was soon sinking with her crew stranded on board. The two other trawlers *Mouleim* and *Mino* were also heavily pounded but stayed afloat. Their crews sought shelter where they could but luckily the *Mouleim*'s galley was empty when a shell smashed right through. The *Mino*, meanwhile was raked from stem to stern with shell holes. Several other craft received hits and the barrage only died down when the Russians perceived a 'larger enemy' bearing down on them. Meanwhile the *Mino* and *Mouleim* along with the *Gull* ran down on the stricken *Crane*. The *Gull*'s boat took off the *Crane*'s crew, both living and dead.

The 'larger enemy' closing on the Russians turned out in fact to be another part of their fleet firing wildly as they approached the action. 'Break off action' was signalled and the firing finally died down about 12.55am, twenty minutes after it had started. The Russians were convinced that torpedo boats were lurking amongst the fleet and they sailed off into the night without making any attempt at assistance.

Vice-Admiral Rozhestevensky's fleet were already well down the English Channel when the *Mouleim*, with her flag at half-mast, led the shell-shattered trawlers back into St Andrew's Dock, Hull. She carried the dead skipper and mate of the *Crane*. By then the news of the disaster had already spread as the injured had been landed by hospital ship. The scene on the waterfront was without parallel in Hull's fishing history[17] and thousands gathered outside the lock-gates as the other shell-splintered vessels including *Mandalay*, *Gull* and *Magpie* tied up behind.[18]

A deputation from the fish trade caught the night mail to London and the next morning went with their MP, Sir Henry Seymour King, to the Foreign Office. By then an outraged British press and people were demanding swift retribution. Crowds gathered in Trafalgar Square and the Russian Ambassador was booed when he left his embassy. Amidst the outrage there was also sympathy for the injured and bereaved. Hull's Lord Mayor received condolences on their behalf from the King's private secretary together with a 200 guinea cheque. The Queen also sent £100 and there were numerous private donations to the swiftly growing disaster fund. One of the most distant notes of sympathy came from the mayor of Tokyo.[19]

Reports of international outrage flooded into London and the incident deepened into a crisis of war-like proportions. Tempers were only inflamed when the Tsar sent a note of regret rather than an apology. By 26 October Britain had twenty-eight battleships at sea ready to attack and destroy the Russian fleet should the Government so order. War seemed imminent for several days, but as time passed passions began to wane. The Tsar apparently ordered units of the fleet involved in the incident to be detained at Vigo whilst responsibility was determined, but Rozhestevensky in fact left only such witnesses as he deemed appropriate. An international commission was set up to investigate and its report, published in February 1905, condemned Rozhestevensky for opening fire and leaving the scene without giving aid to the stricken trawlermen. The British were awarded £65,000.[20] Meanwhile, Rozhestevensky continued his tortuous journey east only to see his fleet finally destroyed by the Japanese under Admiral Togo at Tsu-Shima the following May.

Back in Hull the incident is marked by an imposing memorial and statue of George Smith, skipper of the *Crane*, at the junction of the Boulevard and Hessle Road. This was unveiled in front of a large crowd on 30 August 1906. For the trawler crews life went on much the same. Even when the crisis was at its height and all Europe was alive to the threat of war, many of the vessels of the bombarded fleet continued with their unrelenting trawl and did not return to port until their voyage had run its usual six- to eight-weeks' course. Many more lives had been lost through storm and shipwreck but the trawlers had worked on: earning a living came first.

By this time Hull was the only port still fleeting. Though Grimsby had returned to winter fleeting at the end of the 1880s it had finally abandoned the entire system in 1901 when the number of smacks left working from the port fell below fifty.[21] In contrast, Hull adapted the system to suit steam trawlers and still had the Red Cross as well as the ill-fated Gamecock Fleet. Although steam had replaced sail, life for the fleet fishermen was just as hard. Their craft sailed on six- to eight-week voyages and left the Humber loaded to the gunwhales with coal for the trip. Though steam winches had long replaced the back-breaking job of hauling the net by hand, fish still had to be ferried to the cutter by open boat. Conditions on board remained spartan—it was almost impossible to keep any clothes dry in rough weather and the diet of beef and pudding was still supplemented by weevil-ridden ship's biscuit.[22] Basic toilet functions were carried on over the side for many years to come.

Ashore, the communities in which the trawlermen lived reflected the nature of the trade under limited liability steam fishing companies. Back in the 1850s trawlermen and smackowners had usually lived cheek by jowl. Most of the latter had risen through the trade by way of apprenticeship and the social gap between owner and deckhand—who might well soon own his own smack—was not great. By the second decade of the twentieth century, however, the middle- and distant-water steam trawling companies were increasingly in the hands of land-based capitalists. Though there were still a number of the older generation of self-made former smackmen, like James Alward of Grimsby, their numbers were being depleted by time and their sons and grandsons were taking charge. By now there was a much greater divide between owners and seafarers and the former usually lived in the better areas of the port, if not in the surrounding leafy suburbs and villages; in contrast, the trawlermen lived mostly in crowded, vibrant communities close to the docks in terraced houses off thoroughfares such as Freeman Street in Grimsby and Hessle Road in Hull. Few fishermen ventured far from such streets during their short sojourns ashore and the self-contained nature of these communities was reflected in the number of pubs, clubs, music halls, men's outfitters and pawnbrokers to be found there. Indeed, Hessle Road was steeped in the business of trawling. When vessels had landed in or near the Town Docks most of the early smackmen lived, when on shore, in either the Old Town or the adjacent district of Myton. As the fishing trade moved westward, with first the construction of Albert and then St Andrew's Dock, so the fishermen and their families had followed. Crowded streets of terraced houses were built on what had been the green fields bordering the Hessle Road which ran parallel with the Humber and the docks. Alongside St Andrews Dock was the fish market and across the way the trawler owners' premises. Marine engineering firms were to be found nearby and a dock extension dealt with vast imports of Norwegian herring during the winter. Smokehouses and fish merchants' premises were constructed amongst board schools and family dwellings. Churches and chapels jostled with pubs, clubs, pawnbrokers and all manner of shops. Close by were the Dairycoates locomotive sheds and the railway yards where fish was loaded for inland markets and where, on still and misty nights, the noise of shunting steam trains mingled with the sound of foghorns and bellbuoys from the river. Here, trawlermen's families lived out their lives amongst railwaymen, dockers, bobbers, general labourers, fish process-workers and marine engineering tradesmen. The area was dominated by the docks and the fishing and merchant vessels which sailed from them. At night the main, broad thoroughfare seemed almost to take on the appearance of a western gold-rush town, teeming with life and light. During the day children played in the streets whilst women could be seen standing on the doorsteps gossiping to each other. Trawlermen usually rented small terraced houses—known as sham fours—with at most a tiny front garden and a privy in the backyard. Many families regularly flitted from terrace to terrace and landlord to landlord. Nearer the town centre was the Boulevard and Coltman Street where many of the better off had fine houses, horses and servants. To live here was a mark of status, but before many more decades were to pass such grand streets would be forsaken for the delights—and greater distance—of surrounding villages like Kirkella and Westella, and these large houses would be the first parts of Hessle Road to slip into a decline.

The Hessle Road community epitomized the British trawling trade on the eve of the First World War. It was far different from older fishing communities at places such as Staithes, Scarborough and Brixham. Not only was it a comparatively modern industrial settlement but its people had been drawn together from far and wide. Some families had originated in the older trawling ports such as Brixham and Ramsgate but many more were first- or second-generation seaferers who had been attracted to the port from far and wide—perhaps from other parts of Yorkshire, or maybe Ireland or the Continent. Any number of fathers or—maybe more correctly by then—grandfathers had first been indentured to the trawling trade from workhouses and reformatories across the length and breadth of Britain. More recently, trawling had attracted recruits from amongst the sons of families involved in other Hessle Road trades.

In many respects, Hessle Road was typical of any number of English working-class urban and industrial communities created in the North by the industrial revolution. And yet at the same time it was very different. Hessle Road, and trawling communities like it, were largely based on sea fishing. The grey skies and the broad, brown mud estuary, the sound of trawlers being riveted on slipways or in dry dock, the exuberance of the fish-house girls leaving work arm in arm after being paid, the smoke wafting from the chimneys of the curing houses and, above all, the strong smell of the fishmeal factory, were reflected in every aspect of its character. Though outwardly a modern industrial community, its very heart, and with it the outlook of most of its fishermen and ancillary workers, was still governed by the uncertain hunt for a livelihood on cold, grey seas far from the Humber.

10

The Great War

When war was declared on 4 August 1914 the fish trade was still hard at work. Hundreds of steam trawlers were scattered either singly or in fleets across the North Sea grounds and far beyond. No less than 220 smacks from Lowestoft were out trawling between the Dowsing and the Gab. The summer herring fishery was in full swing as fleets of steam drifters worked the herring shoaling off the north-east coast of England. Virtually every fishing port around the British Isles had vessels at sea. Even in the last days of peace, few people had really believed that a war involving the United Kingdom was imminent even though it was apparent to almost everyone that Europe was in crisis. The sudden declaration took most people by surprise. The Hull Fishing Vessel Owners' Association did not even ask the Fishery Board about the measures being taken to cover fishing vessels in the event of war until July when the storm was already breaking across Europe. At 8.15 pm on 4 August the Admiralty issued instructions, which were relayed on to harbour masters in every east-coast port, that no fishing boats were to be allowed to sail for North Sea grounds and all vessels at sea should be ordered by wireless to make port by dawn.[1]

With fishing boats spread far and wide across the seas, such peremptory orders could not easily be executed. Many vessels did not realize that war had broken out; few possessed wirelesses so messages had to be taken to sea. At Lowestoft a steam trawler set out for the smacks carrying orders for their return. Meanwhile, Aberdeen interned two German vessels, a trawler and drifter, which unwittingly came into the port with good catches.[2] Gradually, the North Sea fishing fleets tied up, the trade ground to a halt and

Plate 4. Early Twentieth Century. The Grimsby Trawler Fittonia.
The Fittonia *was a Casualty of the Great War*

supplies of fish dried up. Those employed in the fish trade seemed more immediately threatened by unemployment than war.

Great Britain, of course, depended on shipping for supplies of foodstuffs and industrial raw materials as well as for carrying manufactured goods around the world. International trade was the lifeblood of the nation and the maintenance of the shipping lanes was vital for survival in any long-term conflict. But whilst it was crucial to make the best use of the available naval resources, in terms of ships and manpower, it was equally necessary to maximize national food production. Yet the joint aims of trying to maintain naval security whilst maximizing fish supplies were to a large extent incompatible. Unregulated fishing vessels were hard to protect and it was believed they could provide a convenient disguise for the infiltration of enemy agents and saboteurs. Consequently the Admiralty's initial thought was to keep all fishing vessels off the North Sea for the duration of the war. But fishing vessels and their crews—initially steam trawlers but then later drifters—were needed as minesweepers to keep the sea lanes open. These formed the Trawler Section of the Royal Naval Reserve but altogether this accounted for no more than 150 vessels and crews.

After the first shockwaves passed, most of the other crews and vessel owners were anxious to get their boats back to sea. The Department of Agriculture and Fisheries prevailed upon the Admiralty to modify their policy and fishing vessels were soon back on the grounds whilst a complex web of naval regulations, based on fishing permits, was gradually evolved to keep their movements under control.

Fishing proved good for the remainder of the summer and all crews benefited from a rise in prices; but the cost of continuing operations soon became apparent. Two Grimsby vessels, *Capricornus* and *St Cuthbert*, were sunk in torpedo boat attacks off Spurn on 25 August[3] and in early September another Grimsby trawler, *Fittonia*, struck a mine and sank in much the same area.[4] A few days later the Hull trawler, *Imperialist* met the same fate some forty miles ENE of Tynemouth.[5] Attacks became almost commonplace. On 16 December 1914 ports and shipping on the east coast were thrown into confusion by German surface raiders. Vice-Admiral Hipper's battle-cruisers slipped across the North Sea to bombard the towns of Hartlepool, Scarborough and Whitby. No less than 120 people were killed and a further 400 were wounded in the attack while a dense minefield was laid inshore between Scarborough and Filey by the 4,350 ton *Kolberg*.[6] At least seven trawlers, and three merchant ships, which were either fishing or on minesweeping duties, were sunk by the *Kolberg*'s mines during the following few months.

Other hazards faced fishing fleets during the War. U-boats used more than mines and torpedoes. On 6 May 1915 the Hull trawler, *Merrie Islington*, with a Scarborough crew, was at work about six miles north-east of Whitby when a U-boat surfaced nearby. Its commander ordered the *Merrie Islington*'s crew to abandon ship in their boat; they later reached the shore safely. The Germans then planted a bomb on board and sank the unfortunate trawler.[7] There were to be many similar attacks and these were not limited to the Yorkshire coast: soon all fishing vessels realized they were liable to receive the same treatment. Many Lowestoft smacks were sent to the bottom in just such a fashion. A further two Scarborough trawlers, the *Florence* and *Dalhousie*, were sunk in this manner on 13 July 1916[8] and on the following 25 September a single U-boat stopped and blew up no less than eleven Scarborough trawlers.[9] Though the crews were allowed to row ashore, their livelihoods had disappeared. A further six trawlers working nearby met a similar fate on the same day. On other occasions U-boats used their guns to sink trawlers and not all crews were so lucky: some were taken back to Germany and held prisoner for the duration. But treatment varied; when the Hull trawler *Hector* was stopped by a U-boat on 3 May 1915 and the crew ordered to leave the boat, the U-boat commander went below deck, took their unfinished meal and a kettle of tea from the stove, wrapped them in oilskins and placed them into their open boat. The *Hector* was then sunk by gunfire and her crew soon met up with the crews of two other boats in the locality which had been sent to the bottom on the same day. The provisions helped all survive an arduous ordeal on the North Sea. Incidentally, the *Hector*'s skipper, Herbert Johnson had a second voyage in an open boat after another trawler under his command, the *Casio* was destroyed by a U-boat's gunfire off the Orkneys less than three months later. Undaunted, Johnson went on fishing throughout the war.[10] Later, as the war at sea became increasingly bitter following the introduction of 'unrestricted' submarine warfare by the Germans, there were also reports of crews being left on U-boat decks to drown when they submerged.

Trawlers were particularly at risk in the North Sea and English Channel but were liable to be attacked almost anywhere. The Hull trawler *Quair* disappeared without trace on an Icelandic trip sometime around 3 November 1916 and was believed to have been sunk by either mine or torpedo.[11] On other occasions vessels were attacked by Zeppelins or aircraft.

The Germans, however, did not come out on top in all encounters with unarmed fishing vessels. On 18 January 1916 the Lowestoft smack *Acacia* was fishing in the company of others when a surfacing U-boat opened fire with a machine gun causing considerable damage to her rigging and boats. *Acacia*'s skipper, J. Crooks, had already been on a vessel sunk by the Germans the previous August and was determined not to be taken again. As the U-boat neared, he waited for a favourable moment then cut the trawl warp and tried to ram her. He missed her by only a few feet and, taking advantage of a fine sailing breeze, ran for Lowestoft and freedom. On another occasion a U-boat, deliberately attacked in a similar manner by an unarmed smack in broad daylight, submerged and was seen no more.[12]

The Admiralty had long been aware that the sea mine posed a particularly dangerous threat to the sea lanes in any war. Many minesweepers would be required and steam trawlers, built for towing and keeping the sea in all weathers, seemed admirably suited, as did their crews. In 1907 the Government had accepted Lord Charles Beresford's recommendation that, in the event of war, steam trawlers and their crews should be used for minesweeping duties. Consequently, in 1911 negotiations were concluded between the Admiralty and certain leading trawler owners as a result of which the Trawler Section of the Royal Naval Reserve was established. Vessels belonging to owners who agreed to the Admiralty's terms were then entered on a special list and became liable for special service as soon as possible after any Admiralty request. During the period of any such service the vessels would be hired from their owners. The hull and outfit of each vessel was to be valued at the rate of £18 for each ton of gross tonnage and £40 per unit of nominal horsepower; the value thus fixed to be depreciated at the rate of four per cent for each complete year of the vessel's age. The Admiralty was entitled to make periodic inspections and could strike vessels from the list. Both parties could also agree to change or substitute vessels on the list and agreements between owners and Admiralty could be terminated by either side subject to six months notice. By the outbreak of war in 1914, 146 vessels were covered by this agreement.

It was soon apparent, however, that many more minesweepers would be required, not only to replace the all too regular losses but also because of the sheer volume of minesweeping and patrol work the Royal Navy was called on to do. At an early stage of the war voluntary hiring gave way to requisition. The mobilization of steam drifters, which were less powerful than trawlers and not fitted out for towing, had not originally been contemplated. Many had been laid up after the suspension of herring fishing but as it proved possible to adapt them for naval service they, like other smaller steam vessels, were also requisitioned. Nevertheless, the Navy preferred the latest deep-sea steam trawlers and would have rapidly requisitioned most of the Hull and Grimsby fleets if left to its own devices. The trawler owners were naturally alarmed by this threat, particularly when fishing was so profitable, but eventually the Ministry of Agriculture and Fisheries and the Admiralty arrived at a compromise solution in which the programme of requisition was spread across many ports and vessels. Even so by the summer of 1918 Hull's trawling fleet was so depleted by requisitions and transfers that the port had virtually ceased to function as a fishing station. In all, some 1,467 steam trawlers, 1,502 steam drifters and 42 motor drifters were requisitioned for the Naval Service during the war.[13]

Utilization of fishing vessels was only one aspect of the problem; manpower was another. Trawlermen in particular were seafarers with skills that the Royal Navy both admired and required. At the outbreak of war a large number of fishermen from all ranks were called up for service as members of the Royal Naval Reserve. Others, who were members of the Trawler Reserve [RNR(T)] were summoned at once to man the Reserve trawlers and many other crews volunteered for service when their vessels were requisitioned. Some trawler owners, alarmed by the Royal Navy's appetite for trawlers and trawlermen, claimed that it could make do with less vessels and crews if they were worked harder. The Admiralty countered that men needed more rest when constantly under the threat of action.

In the event, the opportunity to make the most effective use of the fishing industry's finite pool of skilled labour was sadly missed, partly through inadequate official forethought and partly through the air of patriotic euphoria which followed the declaration of war. The war was initially seen as primarily a short-term land conflict and the call for volunteers to stem the Kaiser's march into France led to a mass mobilization of civilians. But it was carried through largely in terms of numbers and without proper regard for the effective utilization of the recruit's existing skills. Consequently, in the early months, many fishermen were lost to both their trade and the navy by joining the army. William Wells, a Hull trawlerman, was typical of many. A young man with several years valuable experience of working on the North Sea and Icelandic grounds, he joined the army and served throughout the war, yet the nearest he got to naval operations was as a soldier put ashore in the Gallipoli landings.[14] The lack of experienced mariners was keenly felt later in the war, but it apparently proved impossible to locate and release them from the army. Other fishermen, who had long been members of the ordinary Royal Naval Reserve, were swallowed up by the regular navy whilst the Auxiliary Patrol, which was to oversee minesweeping operations, was still in its infancy. Later on, when conscription was introduced, the Board of Agriculture and Fisheries was able to secure a degree of agreement that no fishermen who was willing to serve in the Auxiliary Patrol, if and when needed, should be taken for the army.

Life in the Auxiliary Patrol on minesweeping and patrol duties was hazardous in the extreme. Throughout the war period 214 minesweepers were lost—roughly one a week—and on average each time one went down, half the crew went with her. George Robinson, a second hand from Hull, who had sailed with the Great Northern Steamship Fishing Company before the war joined the Patrol Service in 1915 and spent over two years minesweeping on the paddle steamer HMS *Bourne*—a converted pleasure steamer known as the *Bournemouth Queen* in peacetime. Robinson's diary reveals the stark mixture of tension and monotonous drudgery interspersed with spells of frenetic activity and acute danger that was the lot of those who served in the Patrol Service. HMS *Bourne* was based at Sheerness for much of 1916 and 1917 and usually started sweeping with the rest of her section as early as four or five o'clock in the morning and might work through until early evening, being sometimes called out again to take up a position countering a night Zeppelin or aircraft attack. On occasions, she was involved in actions against submarines or on the work of salvage from wrecks. Though the vessel might work for days, even weeks, on end without encountering a mine, their ominous proximity was all too obvious. The *Bourne* survived the war—and indeed the next one—but vessels and crews all around her went to the bottom with grim regularity. 'You never knew when your time was coming', Robinson remarked in October 1917, after hearing that his old skipper, with whom he had been drinking only the night before, had been blown up on the Grimsby patrol trawler *Strymon* in the Ship Wash.[15]

The fishermen's background, experience and outlook bred an attitude far different from that which the traditional navy expected and the Sea Lords found them difficult to weigh up. The skippers originally recruited into the reserve were seamen of the highest order, used to danger and clearly leaders of men. They were obviously not ratings but hardly from the same class or background as the Dartmouth- or Osborne-trained officers. To settle their position they were accorded the rank of Chief Skipper. The deckhands of the Auxiliary Patrol also exhibited an attitude and approach to the service which differed greatly to that of the regular ratings and petty officers. They were intensely independent, cared little for the niceties of uniform and traditional service discipline, but few could fault them in action. Eventually, the Auxiliary Patrol was almost left as a law unto itself.

Though fishing was subject to close Admiralty regulation, it proved at first impossible to provide vessels with either armed protection or even with their own guns and this enabled the U-boats to wreak havoc. In 1917, however, it was at last possible to fit out the fishing vessels with weapons, and that summer a new naval reserve, known as the Special Fishing Reserve, was formed. Trawlers were expected to fish together in groups under naval command. A number of these vessels were armed and one was fitted with a wireless. The trawlers were commissioned and flew the White Ensign whilst their crews were enrolled in the Reserve. All vessels were placed under the command of the Senior Naval Officer of their base port precisely as if they were units of the Auxiliary Patrol.

The first armaments available for the trawlers were too light and in July 1917 a group of eight vessels outward bound for the Icelandic grounds were destroyed by submarine. When it proved possible to provide heavier guns the trawlermen turned them to good account. At just after 5am on the morning of 20 June 1918 a convoy of trawlers returning from Iceland were attacked by an enemy submarine. The U-boat had surfaced some 7,000 yards to the east and, as the convoy took up line-ahead formation, it began steering on a parallel course, concentrating the fire from its three guns on the leading vessel, the armed trawler *Conan Doyle*. The *Conan Doyle* replied in kind and after about half an hour it made a direct hit forward, but the enemy, though temporarily ceasing fire, made more speed and began to close. Soon the submarine began pouring rapid fire on the *Conan Doyle* until about 7am when contact was lost in a heavy rainstorm. Once the weather cleared, the submarine resumed its attack on the *Conan Doyle*, this time with shrapnel. The British vessel continued firing back and altered the convoy's course towards south to make best use of the wind. The trawlers began to make smoke as the submarine turned its attention increasingly on the other vessels and at about 8am the *Aisne* received several direct hits. By 8.45 am the convoy's situation was getting desperate as ammunition began to run out; the *Conan Doyle* was down to her last fifteen rounds. When all seemed lost the embattled *Aisne* scored a direct hit on her adversary's after-part. The *Conan Doyle*'s skipper, Lieutenant J. A. MacCabe RNVR, had made ready to ram, but with her last few rounds she scored two direct hits, the first of which knocked out a gun and the second went home below the conning tower. The enemy was at once enveloped in a cloud of smoke and steam, and when it cleared about forty-five seconds later, had disappeared without trace. The victorious convoy turned for home once more.[16]

Though badly damaged, the *Aisne* limped into port and was back in the thick of action less than a month later when several trawlers took on another U-boat. Once more the trawlers got the better of the Germans who submerged and made off as about five armed trawlers closed in for the kill. It was clear that armed fishing vessels, working alone or in groups, were no longer such easy targets.

The Admiralty also acceded to requests from the smackmen for arms for their craft and several epic actions were fought with U-boats even though those were a lifetime apart in terms of technology. On 1 February 1917, the day that marked the beginning of the German's 'unrestricted' submarine warfare, designed to bring Britain to her knees, two Lowestoft smacks, the *I'll Try* and *Boy Alfred*, were fishing not far off the Norfolk coast. Like many other smacks, their names were aliases which were frequently changed to confuse the Germans. One of these smacks had been fitted with an auxiliary motor but, more importantly, both were carrying concealed armaments, manned by naval ratings, and trawling in an area known to be frequented by U-boats. As they were hauling their trawls in, at about 12.45 pm they spotted two submarines some five miles distant but heading towards them. The smacks coolly continued to haul in and stow away their trawls but readied their guns. As the first U-boat approached, a figure on the conning tower opened fire with a rifle and called for the *Boy Alfred*'s crew to abandon ship. With the enemy no more than 150 yards off, Skipper Wharton ordered the gun to be unshipped and opened fire, hitting the submarine with the third shot just in front of the conning tower; the craft went down amid masses of steam and water on an even keel. The other submarine also engaged in the action but may well have been sent to the bottom with a well-aimed shot from the *I'll Try* which struck where the conning tower met the deck.[17]

The *I'll Try*, later renamed *Nelson*, and her skipper, Tom Crisp, took on another submarine the following August but this time the tables were turned. Crisp's smack put up a valiant fight with its single thirteen pounder but was outranged by the submarine's larger gun. Skipper Crisp was mortally wounded and posthumously awarded one of the two fishing VCs of the war. This was also the only occasion in which a father and son have been present in an action leading to the award, for the vessel's seventeen-year-old mate was Tom Crisp junior who took over command from his dying father and brought the crew to safety in the smack's small boat.[18] Such actions made some U-boat commanders more wary of taking on fishing vessels on the surface. True, torpedoes could still be used but they were expensive and only a limited number could be carried to sea.

The war at sea became much more bitter as German naval commanders came to regard armed fishing vessels in the same light as naval combatants. Nevertheless, the strategy of arming fishing vessels began to prove its worth: whereas in the third year of the war some 156 trawlers had been lost to U-boats only four were sunk in the fourth. However, the sea mines continued to take their toll and this indiscriminate weapon was to take many fishing vessels and their crews to the bottom long after the armistice of 11 November 1918.

Meanwhile, trawlers and other fishing vessels continued to make landings at many quays and harbours around the British coasts. The U-boat offensive, together with the requisition of vessels and conscriptions of crews for naval duties, had a considerable effect on all levels of the fishing industry. The year 1913 had been the most productive ever recorded for the English fish trade in terms of both landings and value. The total amount of fish landed that year exceeded sixteen million hundredweight and was worth well over £10 million. The war caused a drastic reduction in landings whilst having the opposite effect on values. The average amount and value of all fish landings during the years 1909–1913 had been 14,451,200 cwt and £8,481,400 respectively and though the catch during the years 1915–1918 fell to an annual average of only 4,690,500 cwt its value soared to £9,488,500.[19] At first, the total value of the catch had followed a similar, though less pronounced downward trend to that of landings but by 1916 the rate of inflation was such that the total value began to increase, even though landings continued to fall. Over the next two years a very rapid increase in average value took place, so much so that in 1918, when only a third as much fish was landed as in 1913 its value was nearly half as much again. The average price of fish per hundredweight during 1918 shows an increase of around 500 per cent on the average maintained during the period 1909–1913.

The root cause of this price inflation was the diminution in fish supplies occasioned by the reduction in catching effort allied to the general shortage of foodstuffs. Prices continued to rise until March 1918 when maximum sale prices were introduced by the Government. The average return on each hundredweight landed also reflected a considerable alteration in the relative proportions of each variety of fish landed. During these years demersal fish such as cod, haddock, plaice and sole made up a larger portion of the catch at the expense of cheaper pelagic varieties like herring or mackerel. Indeed, there was a spectacular decline in herring fishing—and therefore drifting—over the war years. The herring curing trade had been effectively detached from its major markets in eastern Europe, and the British (or more particularly the English rather than the Scots), long weaned off the product, could not be induced to eat salted fish no matter how short was the supply of fish.

As a consequence, some areas took a greater interest in trawling. For example, inshore trawling had been banned along much of the Yorkshire and Durham coasts since the early 1890s when the North Eastern District Sea Fisheries Committee had been formed. Inshore fishermen at many of the smaller fishing stations including Staithes and Flamborough had originally welcomed the ban but attitudes began to change during the war. Spurred on by high prices, crew shortages and difficulties in acquiring bait, fishermen from other communities, including Bridlington Quay and Scarborough, pushed for restrictions to be lifted. The Sea Fisheries Committee gradually relented and as they did so more inshore fishing stations took up trawling. By the end of 1918 even the Flamborough and Staithes fishermen, who had been amongst the most implacable opponents of trawling, had taken up the practice and the Flamborough men in particular were said to be 'agreeably surprised' with their landings.[20]

Trawlers traditionally took more fish than linesmen and the conservation of inshore grounds could perhaps be ignored when activity was so curtailed on the offshore grounds. The return from the inshore fisheries during the war enabled many local fishermen to invest in the motorization of their vessels. Prior to 1914 only a handful of the Yorkshire coast fishermen had been able to afford to fit diesel engines to their cobles, but by 1918 many of those operating out of Bridlington Quay and Whitby in particular had been motorized.[21]

Although the risks were greater so were the potential profits. As easly as 1915 it was reported that Scarborough trawlers had sometimes grossed over £200 for one night's catch and earnings continued to grow.[22] The profitability of those steam trawlers still fishing was such that by 1917 some skippers were reported to have earned up to £15,000 in two years despite being entitled to only 10 per cent of the catch's value.[23] Vessels worked whenever they could, sometimes getting back to the job straight after encountering enemy action. On one occasion the crew of a Lowestoft smack rescued two German airmen—an officer and a mechanic—not far from Zeebrugge but decided to remain on the grounds for a further full two days before bringing their prisoners back, in order to make the most of the fishing. The German mechanic lent a hand with the crew whilst his officer provided some amusement by trying to maintain parade ground conduct on the deck of a working smack. Yet despite this attention to work, trawlers saved many other victims of the atrocities of the sea war. William Pillar of the smack *Pillar* showed great courage and seamanship when rescuing seventy-one men from the swamped cutter of the *Formidable* in heavy weather, and one trawler alone saved 166 passengers, mostly women and children, from the torpedoed *Arabic*.

In just over four years of conflict no group of workers contributed a bigger percentage of their number to the war effort than the fishermen. In 1913 the total number of men and boys employed in regular or occasional fishing from the coasts of the United Kingdom was less than 100,000. In England and Wales, including the Isle of Man official figures showed that 37,870 men and boys were regularly employed in all branches of the fishery whilst a further 7,512 found occasional work. Of the number of full-time fishermen, official figures show that no less than 49 per cent were at sometime engaged in naval service and, even taking into account those only occasionally employed in fishing, the total in naval service still reached 41 per cent. Furthermore, of course, a large number of fishermen had joined the army, particularly in the early stages of the war, and virtually all of the remainder who were of military age and fitness were enrolled in a special Naval Reserve to be called up when the Navy required them.

Over the war period some 672 trawlers and drifters were sunk and 416 lives lost through enemy action whilst fishing, and many more were lost on service with the Royal Navy. By the end of the war the Royal Navy had taken some 3,000 fishing craft into service and less than 14,000 men were left fishing from ports in England and Wales of which more than 8,000 were over military age and 400 were below it.[24] Though cut back to the bone by late 1918, the British trade had still managed to contribute an average of around eight million hundredweight of fish to the hard-pressed nation's food supply in every year of the war.

11

Distant-Water Dominance

After the Armistice of 11 November 1918 the aim of the fish trade was to get back to normality as swiftly as possible. Over the following months great attention was paid to the details of demobilizing fishermen and ancillary workers as well as the release of fishing vessels from naval service and the removal, as far as possible, of wartime restrictions on fishermen's maritime movements.

The sheer scale of the trade's commitment to the war effort made demobilization of its resources a complex task. It was important to release fishermen, shore workers and trawlers at much the same pace in order to see that ships were available for the returning crews and that there were sufficient shore staff to deal with the resultant increase in traffic. Not all could be swiftly released as a number of crews had to be retained for minesweeping operations—for the most part volunteers—and vessels had to be stripped of their war equipment before being refitted for fishing. Yet by the end of June 1919 more than 1,000 fishing vessels belonging to English and Welsh ports alone had been sent for refitting.[1]

Many wartime restrictions, including those concerned with landing and lighting, were swiftly removed and early in 1919 the system of vessel permits, by which the Admiralty had controlled the movements of working vessels, was abolished. On 30 August 1919 the Admiralty finally gave permission for fishing to be resumed in all waters except for a few still classified as dangerous.

The war may have been over for the armed forces returning to civilian life and for the German Grand Fleet lying interned at Scapa Flow—prior to being scuttled by its own crews—but the risks of war still afflicted fishermen. Sea mines, part of the undiscriminating mechanism of modern warfare, have remained a deadly hazard ever since. Although minefields were swept soon after the war ended, it was impossible to locate every mine. Floating mines remained a hazard and selected skippers were issued with rifles and ammunition with which to try and blow them up should they be spotted. Other mines were hauled up in nets and skippers were given sets of tools as well as instructions on how to defuse them whilst a naval officer toured the major fishing ports demonstrating methods of making them safe. Not surprisingly, mines continued to send an alarmingly high number of fishing vessels and crews to the bottom well into the 1920s. During the first two years of peace no less than 251 lives and twenty-seven vessels were lost in this way.[2]

War debris also created problems for fishermen. The sea bed around the British coastline was littered with an enormous number of new wrecks; most were at first uncharted and therefore difficult to avoid fouling when trawling. Suggestions that most of these wrecks could be either marked with buoys, blown up, or even salvaged were dismissed as impracticable. Consequently, a great deal of fishing gear was either lost or damaged before fishermen became familiar with their location.

Four years of total war also brought about profound economic and political upheaval, both at home and abroad. The Bolshevik Revolution set Russia on a new course which largely removed it from the orbit of international trade, whilst a nationalistic emphasis amongst many emerging states led them to foster their own economic development, often with the aid of subsidies or tariff walls. Such changes meant that the proportion of international trade to world output would never again reach the levels attained in 1913. For a United Kingdom economy, already weakened by war and yet still heavily reliant on overseas trade but facing growing competition in many of its traditional export markets, the new economic environment was to cause profound problems throughout the inter-war years.

For the fishing industry as a whole, the new political and economic order had its most immediate effect on the herring trade which was deprived of many of its traditional export markets in eastern Europe. The pickle-cured herring trade never truly recovered and output, always so heavily reliant on overseas sales, never again approached the levels attained in 1913. The inter-war period was also traumatic for many sections of the trawling trade. The immediate post-war era was a time of boom, even though the industry had to contend with an unprecedented number of internal labour disputes and was disrupted by railway and coal strikes. North Sea fish stocks were replenished after four years of curtailed fishing and despite the shortage of vessels and crews, landings in 1919 were almost double the wartime average and in 1920 landings of demersal fish in England and Wales set a new all-time record, being some fourteen per cent higher than they had been in 1913. Though the prolonged miners' strike of 1921 disrupted operations, English and Welsh landings exceeded nine million hundredweight in 1922.

Yet the trade was beset with economic problems well before this. All fish prices fell rapidly during 1919 and soon entered into a downward trend that lasted almost to the end of the thirties. Prices fell more rapidly than production costs and profit margins were cut, creating an almost constant pressure to reduce labour costs which aggravated industrial relations. Moreover, the North Sea grounds soon began to show signs of exhaustion once more. The average annual landings of all fish other than herrings from the North Sea by fishermen of all nationalities during the period 1909–1913 had been 434,000 tons. During the period 1928–1932 the level was just 428,000 tons[3] despite the introduction of more sopshisticated gear and equipment. All sectors of the fishing trade suffered. The loss of the herring trade's overseas markets merely exacerbated the situation in many smaller ports and the numbers of English and Welsh fishermen fell from 42,555 in 1919 to 29,011 in 1938. A similar trend was followed north of the border and the number of fishermen fell from 30,762 in 1921 to 21,480 in 1936. The smaller communities which depended on line fishing were squeezed particularly hard by the vicious combination of falling prices and diminishing yields, and in many of them fishing was in terminal decline. By 1935 inshore fishing accounted for about one per cent of white fish landings in England and Wales and around fifteen per cent north of the border. Activity was increasingly concentrated upon trawling and the larger trawling centres in particular. In England the numbers of trawlermen fell much less steeply than did the fishing labour force as a whole, from 22,341 in 1920 to 18,619 in 1938, whilst in Scotland the trawling labour force remained almost static with 3,598 in 1921 and 3,827 in 1936.[4]

Thus most trawling ports were far from prosperous and the trade remained in a depressed state. From 1923, Boston, weakened by the sinking of half its fishing fleet during the war, ceased to rank as a major fishing centre after the port's principal trawling firm, the Boston Deep Sea Fishing and Ice Company, pulled out after a dispute with the harbour commissioners.[5] In many others the construction of new trawlers virtually ceased. At North Shields, for example, no new vessel was acquired between 1919 and 1929, and by 1930 sixteen of the port's fleet of fifty-one vessels ranged from twenty-one to thirty-nine years of age.[6] Even at Grimsby less than twelve per cent of the fleet was under ten years of age by 1934.

Though this was very much the era of the steam trawler some ports still retained their smacks. A survey of fishing vessel registers reveals that Lowestoft still had 130 sailing craft as late as 1926 whilst Brixham and Ramsgate could muster eighty-nine and twenty-five respectively. A few builders still received orders for new smacks from Brixham and Lowestoft fishermen after the war. Indeed, steam trawling never really took root at Brixham. Yet smacks were now of little more than marginal importance to the trade; only 2.6 per cent of the English and Welsh white fish catch[7] was taken by sailing trawler in 1923 and the proportion continued to fall as the number of smacks declined. Motorization, encouraged during the war, continued with some smacks being converted entirely to power whilst others were fitted with an auxiliary engine. Smacks were still being sold off to Scandinavian fishermen whilst others were broken up or converted to yachts. By 1938 there were only twenty sailing smacks left on the Lowestoft register and just five at Brixham. When the war broke out the last Lowestoft smacks were towed up to Lake Lothing, stripped of their spars and rigging, and used to form a barrage against a possible German invasion.[8]

Motorization was one area of modest improvement amongst near-water fleets during these years. The Scottish fifies and zulus—herring vessels—as well as smacks, were fitted with diesel engines and some new motor fishing vessels were actually introduced. Motorization was accompanied by the spread of Danish seining as an alternative to trawling. Danish seining had been devised for catching plaice in Denmark around mid-century. Basically, a

Plate 5. The Fish Quay at Aberdeen—Inter-War

bag-shaped net with long wings on either side of its mouth was set on the sea bed in an open position and hauled towards the boat by means of a winch while the vessel lay at anchor. In the 1920s it became particularly popular along the east coast of Scotland, especially in the Moray Firth area. Here it was adapted to a form known as 'fly dragging' where the net, set in open position, was towed along the sea bed and the wings were gradually brought together as the ropes were winched in. Nonetheless motorization and other methods of white fish capture remained relatively marginal throughout this period. Only seventy-five motor trawlers worked on the near- or middle-water grounds as late as 1937. That year steam trawlers accounted for 97.6 per cent of the white fish catch landed by British fishing vessels in England and Wales.[9]

In Scotland, demersal fish landings were increasingly concentrated on Aberdeen and—although Granton still had a fleet of eighty-three steam trawlers in 1930 with a further nine based on Dundee—by 1934 it accounted for seventy-five per cent of landings by British trawlers in Scotland. But Aberdeen's fleet was also ageing. By 1934 no less than sixty per cent of its vessels were over twenty-years old whilst only eleven per cent had been built in the previous fifteen years.[10] By the outbreak of war the position was little better for no new trawlers had been built for the fleet since 1936.[11]

In the decade prior to the war there had been a vigorous national programme of construction and replacement and few of the major centres had retained vessels over ten years of age. Afterwards many of the former Admiralty trawlers—built during the war for minesweeping duties but with a view to peace-time conversion to fishing—were left by the British trade for overseas buyers to acquire at knockdown prices whilst its own fleet gradually became more obsolete.

The story at Hull was in marked contrast to this catalogue of stagnation. Throughout this period the port concentrated increasingly on distant waters. In 1919 the Hull North Sea Boxing fleets had resumed operations but their tonnage was reduced by about two-thirds. Vessels which had worked in the Hellyer's and Great Northern fleets were converted into single boaters whilst the Red Cross and Gamecock fleets were merged. Whilst other ports continued to concentrate on near- and middle-water grounds the

Hull trawler owners turned increasingly to more distant ones as they began to once more experience declining returns from the North and Irish Seas. From 1936, when Hull finally abandoned the old North Sea fleeting system, it was almost exclusively a distant-water port.[12] Unlike the other British trawling centres, Hull maintained a vigorous programme of investment in new vessels. By 1934 nearly half of its trawling fleet was less than ten years old.[13]

Though traditionally a trawling port, Hull was also responsible for pioneering a number of other innovations. In 1926 the firm of Hellyer's fitted out a former 8,000-ton refrigerated meat transport vessel, the *Arctic Queen*, to process and freeze large quantities of halibut taken from the Davis Straights and Greenland coast. The *Arctic Queen* was joined by the *Arctic Prince* and the two acted as mother ships to a large number of small boats or dories—whose crews were mainly recruited at Aalesund in Norway—which took the halibut by lining. A regular service of carriers then conveyed fresh fish to Hull during the summer, while in winter the two vessels laid up in port and released their frozen cargo on to the market over several months. The *Arctic Queen* and *Arctic Prince*, together with the *Northland* from Grimsby, can be regarded as the forerunners of British factory fishing ships and continued to operate until 1934.

From 1926 Hull also pursued the system of fish filleting with far greater vigour than other ports. Filleting meant that the head, bone and skins were removed at the port whereas previously fish had been transported inland whole. Filleting had been initially encouraged during the Great War for it brought great savings in transport costs: approximately one kit (10 stones) of fish—cod or haddock—yielded 5 stones of fillets.[14] Furthermore, the waste was no longer scattered throughout the country in quantities too uneconomical to collect, and Hull was able to make money by sending it to the fish meal and oil companies which previously had to rely on curing-house residues and unsaleable fish. By the late 1920s the Hull Fish Meal and Oil Company was operating three large factories and could deal with about 80,000 tons of offal per year. The value of this by-product, together with the saving of freight costs, more than balanced the increased labour costs associated with filleting.

Filleting made the coarser distant-water fish more attractive and brought the port an increasing share of the growing fried-fish market: by 1936 the frying trade accounted for more than fifty per cent of British white fish landings. Fillets were especially appreciated by the fish friers who were spreading from their northern heartlands throughout the Midlands and south. Other ports, slower to take up filleting, lost ground in this market and North Shields found Hull fillets strong competition even in Durham and Northumberland.[15] By 1929 about 800,000 consignments of filleted fish were being sent inland from Hull. Not all Hull distant-water cod was filleted or indeed destined for the home market.[16] Twenty per cent of the catch was dry cured for export, particularly to South America.

In 1922 Hull had accounted for twenty per cent of English white fish landings but by 1937 the proportion had risen to forty-five per cent. The port could not have fuelled this spectacular rise without continuing its pre-war policy of opening up distant-water grounds and increasing its interest in the Icelandic and White Sea regions.[17] Though an Anglo-Soviet territorial water dispute restricted fishing off the Murmansk coast in the Barents Sea region for several years, Hull trawlers continued their search for new grounds. In the late 1920s an amelioration in the Arctic climate brought about a retreat of the ice and the Hull trawlermen capitalized by pushing yet further north.[18] Spitzbergen, Cape Kanin and Novaya Zemlya were also visited. However, the most important grounds to be opened up in these years were around Bear Island. Trawling began there in 1929 and in the first year Hull landings from there exceeded 314,000 cwt.

Hull's continual development of the distant-water grounds called not only for newer but also more sophisticated trawlers. The increasing use of wireless at sea assisted with managing fleets working on distant-water grounds. Many 140-foot trawlers had been built during and after the First World War and, with few exceptions, there was little basic change in design until 1931. However, as profit margins were squeezed and the number of long-distance voyages increased the Hull owners began to order larger and faster vessels able to make trips in practically all weathers. Whereas a capacity of 2,000 kits of ten stone each had previously been usual in the early 1920s the new generation of vessels could carry double that quantity. The standard 140-foot vessel became obsolete and new 150-foot vessels were built.

Another innovation was the cruiser stern. This had been pioneered in the South Wales ports some time before, but the first cruiser-stern Hull trawler, *Beachflower*, was built by Cochrane & Sons of Selby in 1931. The cruiser stern gave a full half-knot advantage over vessels with the same engine and power and the new trawlers were able to make well over 11 knots. Many engine improvements were adopted during the decade. The *Kingston Coral*, for example, built in 1936, epitomized everything that was modern about the Hull fleet. One hundred and sixty-two feet in length she carried compound engines fitted with Baver Wach turbines and could make 12 knots. Superheaters helped make the *Kingston Coral* very economical for a trawler of that size and little more than 10 tons of coal a day were consumed. Nine of the class were eventually built. By 1936 the latest Hull trawlers had almost a one-knot advantage on craft built a decade earlier.[19]

The rise of Hull and distant-water fishing was reflected in the relative yields of the various grounds. The figures in Table 3 also illustrate the relative decline of British landings from the North Sea and other home waters. In the pre-war period the ratio of quantities obtained from home waters to those of distant waters was roughly 2:1. By 1929 this had changed to 1:1 with a slight preponderance in favour of distant waters. During the 1930s distant-water grounds became ever more important. The resulting increase in supply, however, also meant a change in the relative proportions of fish landed. Rougher varieties such as cod formed a much larger proportion of the catch compared with pre-war days and their share of the landings continued to increase as the Bear Island grounds were opened up. Bear Island proved extremely prolific and in terms of the catch per unit of fishing time it was estimated in 1930 that the region yielded on average 2,369 cwt for the expenditure of 100 hours of fishing, compared with 1,407 cwt from Iceland and just 130 cwt in the North Sea. The return in terms of quantity of fish obtained per day's absence from port was far higher in distant-water fishing, despite the length of voyages, and such returns more than defrayed the costs of steaming to and from the grounds. By 1930—less than two years after being opened up to trawling—Bear Island ranked third amongst the regions with a yield of 1,185,000 cwt.

Table 3. Demersal Fish Landings by British Vessels in England and Wales from Selected Grounds, 1909–1937

Region	*1909–1913*	*1919–1923*	*1924–1928*	*1929–1933*	*1934–1937**
English Channel	190,000	224,000	191,000	155,000	127,000
North Sea	3,783,000	3,915,000	3,023,000	2,665,000	1,833,000
Faroe	615,000	366,000	763,000	804,000	820,000
Iceland	1,159,000	1,633,000	2,516,000	3,337,000	3,509,000
Barents Sea	273,000	131,000	277,000	568,000	1,331,000
Bear Island/Spitzbergen	—	—	—	758,000	1,627,000
Greenland	—	—	—	112,000	89,000
Norwegian Coast	—	—	—	44,000	355,000

*Four year period only.
Source: Sea Fisheries Statistical Tables.

But the Hull trawlers had not completed their search for new grounds and in 1929 they began working the Lofoten Islands spring fisheries. Other ports generally lacked the drive to find new waters, though Fleetwood owners had discovered new hake grounds on the Atlantic side of the ridge bridging the Faroe–Shetland Channel after fitting out the ST *Florence Brierley* for two exploratory voyages in February and March 1927.[20]

The contrast in performance between Hull and the other centres is clearly outlined in Table 4. Clearly Hull fuelled the inter-war expansion whilst most other ports stagnated.

There is no simple explanation why other trawling centres did not try to emulate Hull's approach during these years. It certainly had little to do with any fear that distant-water grounds would be lost through any wholesale extension of territorial limits by other nations, for such a possibility was not taken too seriously by Britain at that time. In the case of Aberdeen and the other Scottish centres part of the problem was that though they lay closer to the distant-water grounds they were further from the principal centres of English population than the Humber ports and it was less economic to send cheaper distant-water varieties by long-distance overland transport. In fact, though Aberdeen did take relatively large landings of Arctic Cod—largely from visiting foreign trawlers—these were not usually destined for home markets but dry cured for export to the South American market.

Table 4. Landings of Wet Fish (Demersal) in England and Wales 1921–1937 (in cwt)

Year	*Hull*	*Nationally exc. Hull*	*Total*
1921	—	—	7,867,000
1922	1,812,300	7,194,704	9,007,004
1923	1,784,887	6,337,083	8,121,970
1924	2,150,162	6,419,108	8,569,270
1925	2,312,346	6,975,058	9,287,404
1926	2,351,885	6,205,061	8,556,946
1927	2,602,355	6,705,613	9,307,968
1928	2,604,791	6,411,829	9,016,620
1929	3,013,171	6,719,099	9,732,270
1930	4,108,593	7,345,532	11,454,125
1931	4,155,994	6,953,479	11,109,473
1932	4,246,577	6,879,150	11,143,727
1933	4,332,476	6,608,802	10,941,278
1934	4,763,364	6,157,026	10,920,390
1935	5,304,348	6,092,296	11,396,644
1936	6,200,314	6,437,259	12,637,573
1937	6,435,582	6,897,760	13,897,760

Source: Sea Fisheries Statistical Tables 1913–1938.

It has been suggested that owners at some trawling centres lacked enterprise and foresight—this accusation has been levelled at Aberdeen owners in particular.[21] If this was so then it may have been in part a reaction to the profitability of the industry before and during the 1914–18 war. It may also have been the case that the generations which had opened up and developed the trade had been replaced by another, more content to live on the wealth accumulated by the enterprise and hard work of its forefathers. However, such explanations do not fully account for the continual rise of Hull, although it seems likely that the way the trade was organized there fostered a more enterprising environment.

Though steam-trawling operations were generally conducted on a larger scale than was common in other sectors of the British fish trade, the typical trawling firm was still small by inter-war industrial standards—given the level of capital required to prevent obsolescence. In short, many could no longer afford to modernize their fleets as steam trawlers continued to increase in size, sophistication and price. In Aberdeen, for example, the fleet was owned by a large number of small concerns. Although there were a few big companies, the largest of which, with twenty-six vessels by 1939, was the Aberdeen Steam Trawling and Fishing Company (a subsidiary of MacFisheries and the Leverhulme Group) most were much smaller. A number were owned by a single individual, sometimes the skipper, or else a small partnership held shares in a single vessel. In 1934 no less than forty-one per cent of Aberdeen's fleet was owned by concerns operating less than five trawlers and almost seventy per cent was held by firms with no more than nine vessels. This situation had changed little by the outbreak of the Second World War.[22] The structure of ownership was diverse and though seventy-five per cent were limited liability companies a number of vessels were still split between individuals holding sixty-fourth shares, a relic from the old sailing ship days. Many shares were held by outside interests who were often considered by insiders to bring neither credit nor stability to the trade by being anxious to participate in good times but keen to transfer their capital elsewhere when prospects were worse.

At Fleetwood there was a similar proliferation of small owners. In 1939 its fleet of 112 vessels was split between no fewer than forty-one companies or individuals but many craft were operated by management units on their owner's behalf. Even so, twenty-one separate management units operated the 112 trawlers.[23]

The distant-water trade upon which Hull increasingly concentrated was equally open to its neighbour Grimsby but, though it had developed and maintained a deep-sea sector, the Lincolnshire port continued to concentrate on taking prime fish from middle- and near-water grounds. Indeed, though Grimsby landings by 1937 were much lower than Hull's, they still commanded the greater value, thus stimulating vigorous argument about which was the world's premier fishing port. Like Aberdeen and Fleetwood, most Grimsby trawler companies were also relatively small scale. Only five firms had a workforce above 250 whilst just six others employed more than 100, and some of these were fish merchants as well as trawler owners; in 1934 two-thirds of the Grimsby fleet was owned by concerns with fewer than nineteen vessels apiece. There was also a tendency in some firms for ownership to be divided and sub-divided by successive generations and for increasing outside claims to be made on the profits, thus depriving these firms of the capital resources essential for replacement and expansion.

In the later thirties Grimsby's decline was halted: the opening of the new No 3 dock in 1935 provided the trade with modern marine engineering facilities and three hydraulic coal-hoists to speed up bunkering, as well as doubling the water space available. Forty-six new vessels were added to the fleet in the three years 1935–1937 with the lead being taken by a small group of firms including Associated Fisheries Ltd, established back in 1929 with a capital of £500,000.[24] This firm also bought fifteen large, German-built Fleetwood trawlers from MacFisheries who had operated them with little success. The arrival of these vessels in 1938 made a big difference to the port's fish supplies.

In contrast to other ports, Hull's trade was organized on a much greater scale. In 1935 fifty-two per cent of Hull's fleet was owned by companies possessing more than nineteen vessels— compared with thirty-three per cent at Grimsby. Larger firms could more easily raise the capital necessary to modernize and expand the fleet and those in Hull seemed to possess managements more willing to sustain such a policy. Moreover, though there was mutual co-operation at other ports, none had developed it to the level of Hull. Hull trawler owners, fish merchants, curers and wholesalers had financed a whole range of jointly operated concerns, including ice factories, fish meal and manure plants as well as a cod-liver oil refinery and mutual insurance companies. This modern structure, backing up the trawling fleet, was to the benefit of all and contrasted markedly with the situation at some other ports.[25]

Hull's success in distant waters brought the trade new problems. Although 1930 had been a record year for landings, thanks in particular to the new Bear Island grounds, it also saw an eighteen per cent fall in the catch's value. Part of this fall was due to the large increase in lower quality distant-water fish coming on to the market, but 1930 was also a year when all food prices fell rapidly. Landings remained high during 1931 and 1932 and food prices continued to fall as the world economic depression took hold with the result that fishing became increasingly unprofitable. Though fish consumption per head of population steadily rose over the inter-war period it was clear that in the early 1930s landings were outrunning the capacity of the market.[26] Even the prime-fish producers were affected by the influx of the cheaper and coarser Arctic Cod. Only Hull with its modern, highly efficient fleet could make a profit and the trawling industry was soon in disarray.

It was sometimes argued that the trade was concentrating on building the best of ships to catch the worst of fish.[27] The decline in overall quality of fish due to the growing concentration on distant waters was especially marked during the summer months when a combination of long voyages and warm weather caused excessive deterioration. Methods of keeping and distributing fish still left much to be desired: there were complaints about over-long exposure on the fish quay and of railway companies using open wagons covered only with tarpaulin instead of purpose-built fish vans.

Not surprisingly, the distributive side of the trade was in a far from healthy state and the problem of stale fish harmed the trade's image with the consumer. Though the numbers of port wholesalers had considerably increased since 1919 the average volume of fish each handled had declined at many ports including Aberdeen and Grimsby; in contrast, it had risen in Hull. Nor were wholesalers making better profits on this lower throughput for, excluding Hull, net profits were only 0.1 per cent of turnover. Once more Hull, with figures of 2.3 per cent proved the exception. Little capital was required to set up in this line of business and the comparative ease with which new concerns could establish themselves contributed to a general air of instability amongst merchants at many ports.[28]

At Billingsgate, the main wholesale market, merchants dealing as principals in 1934 showed an average net profit of 0.7 per cent on turnover whilst those dealing partly as principals and partly on commission on turnover made 2 per cent. Those dealing only as commission agents generally made a loss of 0.5 per cent during that year. But the average volume of trade transacted by wholesalers in the major inland markets was very much greater than was the case with port wholesalers, and in many provincial markets there was greater stability of numbers with many stances (stall or stand) remaining in the hands of families for long periods of time. On the retail side, the number of fishmongers had apparently declined, with the net profit averaging 2.9 per cent of turnover by 1934, whilst there had been a considerable increase in the number of fish friers.

Perhaps the most disturbing feature of the distributive situation in the early 1930s was the wide and ever-increasing margin between fish prices at either end of the chain. In 1934 the Sea Fish Commission found that unheaded white fish roughly doubled in price between port and consumer and trebled if headed.[29] Despite this it was evident that few in the merchanting sector were making reasonable profits.

The traditional response of the trawling trade at a time of falling fortunes had always been to seek ways of increasing catching efficiency, and this remained the case in the 1920s. Though many of those who worked the near- and middle-water grounds could not or would not acquire new vessels, they continued to improve fishing gear. The original otter trawl rapidly gave way to an improved version known as the Vigneron-Dahl trawl in the later 1920s when it was discovered that catching efficiency was markedly improved by separating the otter boards from the net by means of long wire ropes known as bridles. These bridles dragged along the sea bed, stirring up fish and herding them towards the trawl mouth. Heavy hardwood rollers called bobbins were now also being fitted to the ground rope and this enabled the trawl to be used on rougher, previously unworkable, ground. The bobbins lifted the lower lip of the trawl as they rolled over rocks and stones reducing wear and tear.[30] Catches initially improved everywhere it was adopted but whereas the vast distant-water cod stocks could stand this, the story was much different in near and middle waters. Here there was a temporary increase in catches of haddock, plaice and halibut but the hard-pressed stocks were being stripped to the bone and after 1930 the landings of such species fell to the lowest peace-time levels hitherto seen in the twentieth century.

The first reaction of the trawling trade as the situation worsened in the early thirties was to lobby for duties on imported fish.[31] Fish imports were a convenient scapegoat and a number of Grimsby owners blamed them for their failure to modernize fleets during the 1920s. Some foreign fleets had certainly benefitted from state subsidies and though total imports and landings of foreign-caught fish were about two-thirds higher in 1930 than they had been in 1913 they comprised large amounts of herring as well as canned and frozen fish. Direct landings by foreign vessels of fresh fish formed less than eight per cent of total British landings. Moreover, many foreign landings, of distant-water cod at Aberdeen for example, were then cured and re-exported. At Hull, where the opposition to imports was particularly vociferous, they formed less than two per cent of landings.

In 1932 the Government set up a committee to investigate the trawling trade's problems and its findings led us to the Sea Fishing Industry Act of 1933 and the introduction of several orders, one of which, the Restriction of Fishing in Northern Waters Order (1933), prohibited landings from the distant-water grounds of Bear Island, the Murmansk Coast and White Sea during the months of June, July, August and September. British fishermen were also prohibited, even on the high seas, from using nets with meshes below a prescribed minimum size and it was unlawful to sell any hake, haddock, plaice, soles or dabs which were below a certain size. Restrictions, in the form of quotas and duties, were imposed on landings and imports of foreign-caught fish. In effect, this Act reversed Huxley's dictum enshrined in the 1868 Sea Fisheries Act that fishermen and fishing vessels should be able to fish when, where and how they liked.[32] For the British fisherman it brought to an end sixty-five years of *laissez faire* on the high seas. The 1933 Act also arranged for the constitution of a Sea Fish Commission charged with carrying out an in-depth investigation of the industry from sea to consumer. Sir Andrew Duncan was duly appointed chairman and its first review, into the herring industry which was in an even more parlous condition, was submitted in 1934 and led to the creation of the Herring Industry Board in the following year. In March 1936 the Sea Fish Commission submitted its second report, this time on the white fishing industry, and the 1938 Sea Fish Industry Act followed many of its recommendations. A permanent White Fish Commission was to be set up consisting of three individuals independent of the trade with wide-ranging powers covering a host of aspects from the registration of port wholesalers, the creation of a marketing board, and the encouragement of marketing schemes, to the introduction of special co-operative arrangements for inshore fishermen. The Act also provided for the creation of a White Fish Industry Joint Council to advise the Commission. The outbreak of war in 1939 prevented many of these arrangements being instituted until peace returned, with the exception of the Joint Council upon which sat representatives of trawler owners, merchants, processors, fishmongers and fish friers.

Over-production remained a problem of the deep-water sector during the last years of peace. In 1938 Hull and Grimsby owners devised a voluntary restriction scheme which involved the laying up of twenty per cent of such vessels. The middle- and near-water sectors largely continued to stagnate and though rearmament tended to stimulate the national economy it had the opposite effect on the trawling trade in some ports. The outbreak of war would involve the requisitioning of trawlers and the first vessels the Admiralty would want would be the newest. Hence there was a positive disincentive to modernize, and many old vessels from Aberdeen, Fleetwood and Shields soldiered on into 1939 as the European political situation continued to deteriorate.

In essence, the trawling trade was principally confronted by two different and somewhat contrasting problems during the period under review. Whilst the near- and middle-water grounds were afflicted by over-fishing and stock denudation, the rich catches made on the distant-water grounds continually threatened to overwhelm the market and trade. In reality, the lack of adequate conservation policies was at the root of the troubles. Since at least the 1870s trawling technology had tended to lead to over-fishing and the trade had usually responded by increasing catching efficiency and seeking out new grounds ever further afield. During the inter-war years many trawling centres seemed unable to pursue the latter line. In general they failed to find new grounds for the prime fish they specialized in and their efforts to increase catching efficiency merely exacerbated the long-term problems of stock denudation on the old grounds. Thus much of the trawling trade slipped into unprofitability. For the first time since the 1866 Royal Commission the Government was forced into legislative conservation in international waters. However, Hull—and later, to a lesser extent, Grimsby—continued with the old methods with an undiminished vigour. New distant-water grounds were opened up by larger and ever more efficient trawlers. Though the exploitation of distant-water grounds aggravated problems for the near- and middle-water sectors it did allow the industry to keep pace with—and even overrun—the British and overseas demand for white fish. Moreover, the sheer wealth of northern fish stocks weakened the conservation issue. Concentration on distant waters postponed but did not cancel the trade's day of reckoning. The world's sea fish resources were finite, even in the vast waters of the northern trawl.

12

Inter-War Life and Labour

The trawling trade had, of course, a sporadic history of unionization, even industrial militancy, stretching back, in the case of Hull at least, to the 1850s. Concessions had been won by the men at various ports during the eighteen months prior to the war, but the accompanying unrest had been followed by a period of wartime quiescence when many trawler crews had been drafted into the navy. Following demobilization, however, the trade was caught in the labour unrest then sweeping the country and faced an unprecedented series of disputes lasting into the 1920s. The high fish prices prevailing immediately after the war made many groups of British trawlermen determined to push for better pay, shares and conditions. Though some of the disputes at lesser trawling centres, including Shields, Boston, Granton and Hartlepool, met with marginal success on the men's side, the most important strikes—which took place at Aberdeen and Hull in 1919, Fleetwood in 1920 and Grimsby in 1922, followed by Aberdeen again in 1923— saw capital, despite some concessions, generally get the better of labour.[1] A glimpse at the Hull and Fleetwood disputes helps to explain how the owners retained their hold on the situation.

Back in July 1917 Mr G. W. McKee of the Hull Seamen's Union and Mr Gibbins of the Humber Amalgamated Engineers' and Firemen's Union approached the owners with a claim for increased remuneration for trawler engineers and firemen. Initially, the two unions differed about the form of increase they wanted. Up to this time the engineers had steered clear of the profit-sharing principles which had been spreading through the trawling trade since the Grimsby strike of 1901, and the Hull Seamen's Union had also maintained this line, pushing for a straightforward increase in weekly pay. But wartime inflation and the spiralling value of fish landings and thus trawler profits (fish prices remaining unregulated until March 1918) had encouraged the HAEFU, which represented the larger number of Hull trawler engineers, to take a different tack and press for payment on the profit sharing principle. The trawler owners, faced by conflicting claims, asked the unions to agree a joint approach and the HAEFU's strategy prevailed. The owners soon agreed to an arrangement based partly on a weekly wage and partly on share similar to that already negotiated at Grimsby but, ironically, the men then threw this out at a mass meeting and an increase based on the original Hull Seamen's Union request was swiftly negotiated.[2]

But fixed-wage agreements failed to keep pace with inflation or profits and at the beginning of March 1918 the Engineers' Union was again requesting the introduction of a share element. Both unions and management then concluded an arrangement more or less on the lines thrown out by the men some seven months before under which the share element would be paid on gross earnings, though the owners reserved the right to renegotiate payments on net earnings after the war.[3]

In January 1919 a number of Grimsby trawler crews struck, without the initial backing of any union leadership, in what appears to have been a grass-roots demand for better wages and share levels. There was still a shortage of demobilized trawlermen and concessions were swiftly gained from the port's trawler owners who were anxious not to jeopardize the profits they were making in the post-war boom. During the same month the Hull trawler engineers capitalized on the situation to obtain a further increase in their poundage share. The apparently concessionary mood of the owners was matched by increasing militancy amongst engine room personnel as more men and vessels returned from naval service. On 3 June 1919 the owners were informed by the HAEFU that a general meeting of engineers had passed a resolution demanding two firemen trimmers for all Icelandic and North Sea trawlers and that unless this claim was immediately settled they would not take the trawlers out. The owners considered such an approach to be unconstitutional, claiming that it violated an agreement reached with the unions back in 1917 whereby any changes in conditions would be discussed by a conference of all concerned.[4]

At this point major differences between the unions became obvious. The Hull Seamen's Union, backed by its close associate, the by then moderate National Sailors' and Firemen's Union, also viewed the strike as unconstitutional and called for those who had come out to return to work. The HSU's view was that all claims should be dealt with through the established negotiating machinery. However, though a strong force in the port, the Hull Seamen's Union membership was drawn mainly from the crews of merchantmen. Its Trawler Section was relatively small and its stand was attacked by Gibbins of the rival HAEFU. His union maintained the more militant line as more engineers, never previously members of any union and therefore not directly party to such agreements, came out on strike. It was known that the owners sought to alter the terms of the share agreements in order to make payments on net rather than gross vessel earnings and in response the demands of the workers were increased to include a request for a 2d in the pound net share as against 1½d in the pound gross. On 13 June 1919 the owners agreed to this and also conceded that an additional trimmer could be carried on any trawler over 130 tons.

Up to this point only distant-water trawlers had been involved, but hardly had these concessions been agreed when the men changed their demands: they now requested two trimmers for North Sea vessels as well a continuation of payments on gross or the alternative of going back on to a weekly wage system whereby the Chief Engineers would be paid £8 per week, the Second £7 5s and the trimmers between £2 17s and £3 per week. Faced with continually shifting demands and inter-union disputes the owners chose to dig in and stick to the terms of 13 June.[5] Soon, virtually every trawler in the Hull fleet was laid up.

The strike dragged on into July and a conference called to settle the issue, chaired by Mr Douglas Gray of the Ministry of Labour, collapsed after the HAEFU refused to negotiate alongside the officials of the other two unions. Throughout July the inter-union rivalry grew more bitter with both the HAEFU and the HSU claiming to speak for the engineers. In practice many of the strikers had not previously belonged to any union and the militant stand of the Amalgamated Engineers seemed to win more support from their ranks. On 1 August the union side was plunged into greater confusion when a meeting held by G. W. McKee of the HSU passed a resolution in favour of going back to sea on the terms offered. Gibbins retorted that the strike was not breaking up and his men were staying out.[6]

Though only the engineers were on strike, vessels had been unable to sail without them and hundreds of fishermen returning from the navy were laid up. Once vessels manned by the HSU engineers began fitting out for sea they found little difficulty in recruiting deckhands, some of whom had only scant sympathy for the engineers. Nevertheless, the majority of engineers were still out and on 12 August the owners finally agreed to meet with the HAEFU alone. The terms then offered seem to have changed little from those previously agreed and were rejected by a mass union meeting. Soon more vessels were getting away and within a week Gibbins' supporters were forced to go back on the terms they originally rejected on 13 June. These were:

Iceland and White Sea fishing grounds

Chief Engineer	£4 5s per week and 2d in the £ net
Second	£3 5s per week and 2d in the £ net
Trimmer	£3 0s per week

The terms for the North Sea vessels were the same except the trimmer received only £2 17s per week. All trawlers going to Iceland would ship two permanent trimmers.

Gibbins afterwards claimed that the union had not been defeated and had gained an important point in getting the owners' association to negotiate directly with them, rather than in conjunction with the other two unions.[7]

Unlike Hull, the first ever Fleetwood trawler strike, which began on 6 February 1920, soon won the support of all crew sections. Initially, the deckhands sent out the call for action but the engineers, skippers and mates quickly followed with demands of their own. Unity, however, was only a façade for the men were represented by separate associations and had different aims. The skippers and mates had their own guild and recognition of this body by the owners—as well as an increase in share money—was a primary aim. The recently formed National Union of British Fishermen looked after the deckhands' interests whilst the engineers were represented by the Humber Amalgamated Engineers' and Firemen's Union, so prominent in the Hull dispute.

The owners were able to exploit these differences by settling the deckhands' claim within a week. They agreed to give them 10s a week risk money on top of their existing wages and share money. From then on the factionalism amongst the workforce became more evident. As at Hull, the engineers had once been vehemently opposed to the share remuneration but they too had seen the apparent benefits it could offer at a time of rapidly rising prices and increased incomes and now wished for its inclusion. But such a demand merely widened sectional differences and made further united action less likely as the deckhands felt it was their hard labour which secured the catch out on the grounds whilst the engineers stayed below and drunk tea. The engineers, in contrast, felt that they did most of the work whilst steaming to and from the grounds.

Though most Fleetwood trawlers were tied up by the time the dispute dragged into its sixth week, the strike was continually being weakened by disputes between union and non-union members, as well as members of different unions. Some skippers joined the NUBF and took their trawlers to sea whilst the deckhands reaffirmed their decision to return to work.

After six weeks of increasingly bitter strife the dispute was sent to arbitration and the men returned to work. Two months later the arbitration award gave the trimmers an extra 14s a week and the engineers a structure of remuneration which included 2d in the pound on net earnings whilst the skippers and mates gained nothing. The owners had now succeeded in getting all classes of crew on to the share system.[8]

In these and other disputes a key factor was the trawlermen's lack of unity. Traditionally, as we have seen, trawlermen had been difficult to unionize and the problems of organizing them were little easier in the twentieth century than they had been in the nineteenth. The more ambitious still generally aimed to become mates or skippers, whilst their way of life, with short periods ashore and an increasing number of voyages to ever more distant waters, remained unconducive to long-term organization and planning, as well as rendering the collection of subscriptions extremely difficult. Though vessels and crews were larger than in the days of smacks the workforce was still split into relatively small units on trawlers scattered across the northern waters—such men had little opportunity for swift and concerted action. True, the old problem of indentured labour was virtually a thing of the past but the replacement of sail by steam had introduced a new faction to the crew in the shape of the engine-room personnel. Moreover, a number of trawlermen had long felt disinclined to work closely with other seafarers in general seamen's unions, partly because they felt that their's was the harder way of life, but also because they had little in common with the lives of merchant seamen.[9]

After the end of the war there was little evidence of concerted action between ports—apart from some attempts to spread unionization. However, for some time afterwards occasional disputes, often over similar issues, broke out around the coasts. In many of these the trawlermen's cause was weakened by the lack of a strong union tradition, and their ranks were wracked by factional interests and inter-union disputes which clouded any notion of a common interest or aim. As a result, the men not only exhibited a lack of unity but, as we have seen, sometimes gave the employers' side conflicting or confused messages which hardly encouraged them to deal more closely with the union leaders. The employers, in contrast, had developed strong associations in many ports and in 1919 established the British Trawler Federation (which Hull owners did not join for many years) to oversee their national interests; but they nevertheless remained suspicious and hostile to the general concept of trade unionism amongst the many sections of their workforce.

Trade unionism amongst trawlermen remained weak throughout most of the twenties and thirties. Back in 1917 a retired Grimsby skipper, Captain Bingham, had set up the National Union of British Fishermen and by the end of 1919 it claimed a membership of 5,000, based mainly at the Humber ports but with other active branches at Aberdeen and Fleetwood. The National Sailors' and Firemen's Union—which absorbed the Hull Seamen's Union in 1922—also claimed more than 5,000 members[10] about the same time but its strength was largely concentrated in ports such as Shields, Lowestoft and Ramsgate. Certainly, the numbers of trawlermen in both unions seem to have fluctuated widely in the following few years. Inter-union rivalry grew more bitter and there were frequent accusations of poaching and blacklegging.

In February 1921 Government price controls were lifted and as the post-war boom subsided the owners began to cut labour costs and the divided trawlermen were unable to stop them. In April 1921 a revised scale of wages was pushed through. Skippers and mates in the three largest English trawling ports, Hull, Grimsby and Fleetwood, had their trip money abolished. Their share money was kept fixed at 1⅜ and 1 share respectively, supplemented by the possibility of £100 and £25 annual bonuses, but subject to stringent conditions laid down by the owners. Deckhands saw their regular six-shilling sea bonus disappear to be replaced by an extra penny in the pound share money. The engineers fared little better. Many groups of workers faced wage cuts as profits tumbled and unemployment rose in the early twenties, but few lost as much as the trawlermen. Railwaymen's wages fell from sixty-eight shillings to fifty-six shillings and miners lost two shillings per shift but fought these cutbacks with bitter disputes whilst the divisions amongst the trawlermen at this time made it difficult for them to resist the owners. Fleetwood deckhands and trimmers, for example, proved powerless to prevent their average earnings falling.[11]

It is perhaps not surprising that most trawlermen drifted away from unions in the 1920s. Many of the most ardent remaining trawling trade unionists left the Hull branch of the NSFU in 1926 when it failed to support the General Strike,[12] whilst the NUBF was absorbed by the Transport and General Workers' Union in 1922. By the early 1930s the TGWU could still claim about 500 members at Hull and a small presence at Fleetwood, but in many other ports union organization, of deckhands at least, had all but disintegrated.

Though trawler owners at many ports did not always co-operate as closely as they might on ventures beneficial to the trade they tended to pull together when faced with what they saw as outside interference or labour troubles. Each port had its own owners' association and, of course, by the thirties the British Trawler Federation, formed in 1917, represented most owners with the exception of those at Hull.

Despite making a few initial concessions, trawler owners at most ports now had the whole of their seagoing labour force on the share system. Disputes over the stocker-bait money, liver-oil money and poundage shareout now affected all members of the crew and tended for the most part to increase divisions amongst them. The lack of strong union support meant there was little chance of negotiating better working conditions. Furthermore, though many men believed that corruption was widespread they usually felt powerless to stop it. Owners at some ports were quite prepared to blackball, that is prevent anyone they regarded as a trouble-maker from getting a ship for several months.[13] Though illegal it was often claimed that the ship's runner, who recruited the crew, demanded a backhander from those looking for a ship. Moreover, the men were suspicious of the owners over the share system. Most trawler firms did not clearly state how deductions from gross earnings were arrived at and accounts—prepared by the owners—were rarely supplied in a form that were easy to check. As the Report of the Committee on the Fishing Industry remarked:

> A system payment involving part of the remuneration in the form of shares will only work satisfactorily if the financial results of the voyage on which his share depends are presented to the fishermen in a way which commands his confidence.[14]

Such a situation remained a focus of aggravation which resurfaced again in the 1935 Hull Liver Oil Strike. The dispute broke out when payments made to crews for casks of liver oil were reduced without consultation. Livers had long been regarded as a perquisite of the crew and until 1932 they had been barrelled at sea for the refining company with each barrel or jar fetching 12s 6d for the crew. That year a system of boiling livers at sea was introduced and by 1935 two-thirds of the Hull fleet had gone over to it. In effect the first stage of the refining process was now being carried out by the deckhands at sea and the crew received payments on casks of oil they produced.[15]

The owners were to claim that the initial price of 45s per cask had been somewhat experimental and, given that two and a quarter jars of livers were required to produce one cask of oil, the reduced prices being offered for casks still compared favourably with those the men had received under the old system. Moreover, as it was evident that the market for cod-liver oil was depressed and large stocks had accumulated, so employers and refiners argued that it was necessary to reduce payments in line with current market values. For their part the men contended that the boiling system reduced the refiners' production costs whilst involving the crews in yet more work at sea for which they were not adequately compensated. However, they did not argue so much against a reduction in money as against the arbitrary manner in which the reductions had been imposed.

A subsequent Board of Trade Court of Inquiry found that the dispute turned into a bitter strike largely because of a lack of communication. The owners believed in the 'hard old-fashioned approach', as *The Times* remarked, to handling their men. The Transport and General Workers Union, which had the greatest membership amongst the deckhands of any union, had been unable to obtain access to the owners' association to discuss the dispute. Indeed, the owners did not recognize the TGWU and unlike many other industries the trawling trade had failed to establish, or see the need for, any form of joint negotiating machinery. When the TGWU Assistant General Secretary, Arthur Deakin, called at the offices of the Fishing Vessel Owners in order to introduce himself to the Chairman he got no further than the office boy.[16] The industrial court of inquiry, subsequently convened under Sir James Baille, cut the reduction in liver-oil payments to half of the owners' arbitrary figure.

The Hull Liver Oil Strike represented something of a watershed for trade unionism in the trawling industry and in the following year, at Fleetwood, sectional differences were finally sunk in an amalgamation initiative which led, two years later, to the creation of a single national fishing section within the TGWU. These two ports, now leaders in distant-water fishing, pointed the way towards the permanent unionization of trawlermen, a movement that was to be consolidated after the war.

Throughout the inter-war years fishing and especially trawling remained a hazardous occupation. The lack of union influence hardly encouraged an improvement in conditions. Over the five-year period 1932–1936 an annual average of one out of every 220 fishermen working in first-class fishing vessels was lost whilst many more were seriously injured. Countless others had lucky escapes when vessels were driven ashore on hostile coasts. In some years

Plate 6. Crew Members of the Hull Trawler Nyemetzki

losses were concentrated on particular ports. In 1935, for example, one and a half per cent of Grimsby's fishing workforce was lost over the twelve months. Many vessels were lost far from their home ports but two of the most poignant Hull losses, the *Edgar Wallace* in January 1935, and the *Lady Jeanette* in March 1939, occurred within hailing distance of St Andrews Dock. Both trawlers, returning to port after long voyages, were caught by strong Humber tides and overturned on sandbanks. Of the entire crew of the *Edgar Wallace* only the mate survived by clinging to the mast, whilst seven of the *Lady Jeanette*'s crew were lost in the swirling Humber mud.

Another hazard continually facing some distant-water trawlers was that of running foul of fishery protection vessels. Throughout this period trawlers were continually being arrested by Danish and Norwegian naval vessels for allegedly fishing in territorial waters. Occasionally such occurrences, long part and parcel of fishing life, were blown into international incidents. In January 1925 Skipper Loftus of the Grimsby trawler *Our Alf* was sent to prison in Iceland for 10 years in default of a £750 fine for illegal trawling. The ensuing outcry illuminated the different Icelandic and British attitudes to these disputes. To the Humber fish trade at least, Loftus's sentence was 'utterly alien to all British ideas of justice'.[17] Loftus was apparently known as a 'lucky skipper', that is one who could be relied upon to make good catches, and as such had attracted the attention of a Danish fishery cruiser. The cruiser had first tried to arrest the trawler by sending over a boat-load of armed men, but the crew of the *Our Alf* had repelled them after hitting one junior officer over the head with a bucket. The Hull and Grimsby trade argued that their trawlermen were subject to constant harassment by gunboats and that the Icelandic law which decreed that no vessel should enter territorial waters unless its trawl was stowed away was too harshly applied. They claimed that vessels had been arrested after being forced inshore by violent weather before they had time to pack away their trawls.

Iceland viewed the problem differently. They complained of constant incursions into territorial waters by large foreign trawlers which carried away their inshore fishermen's lighter gear. In order to back up the Danish gunboats a number of Icelandic fishermen had been enrolled as water police and given fast motor boats to chase after such wrecking trawlers, but they complained that the trawlermen often forcibly resisted arrest and on occasion had even sailed for home carrying Icelanders who had tried to board their trawlers. Loftus, it was argued, was a constant thorn in their side. Nonetheless, he only remained in prison until his fine was paid.

Fishing limits are always hard to enforce rigidly and vessels have always been tempted to overstep them. Some skippers, under pressure from owners to keep up catches even when the weather was at its worst, were tempted by the rich inshore grounds. Their life bred the kind of outlook which encouraged them to take risks. The same attitude which encouraged some trawler skippers to ignore Admiralty regulations about fishing in mine-infested waters when good earnings were to be made during the Great War tempted others into Icelandic territorial waters after the return of peace. But such an attitude hardly encouraged the kind of bi-lateral relations that were to be so important after 1945.

However, the trawling trade seemed to see no ambiguity in slating the Danes and Icelanders in one breath for prison sentences 'utterly alien to British ideas of justice' whilst praising them on the next for the help they afforded to shipwrecked vessels. During the same winter that Skipper Loftus languished in gaol, the Hull trawlers *Viscount Allenby* and *Vera* were wrecked off the south coast of Iceland and their crews received great hospitality and assistance from the Icelanders who provided them with food, shelter and horses for their overland journey to Reykjavik and the mail ship home.[18] If poaching pushed the two nations apart then disasters brought them closer again. The year 1925 produced a particularly bad winter in Icelandic waters: British and Icelandic vessels searched for the Icelandic trawler *Liefurheppni* and the Hull trawlers *Field Marshal Lord Robertson* and *Scapa Flow* throughout late February and early March in some of the roughest weather ever recorded. Despite their efforts the vessels were never found and were assumed to have foundered. Altogether, some eighty-five British and Icelandic men were lost. The sea did not recognize nationality or the territorial disputes of men.

13

The Second World War

When war was declared on Germany on Sunday 3 September 1939 the British distant-water trawling fleet was either tied up in port or else making for home, having been recalled by Admiralty order on 27 August. As in the previous conflict the Government recognized that trawlers and fishermen had an essential role to play in the maintenance of Britain's sea lanes and the Admiralty was keen to get minesweeping operations underway.

In the summer of 1939 the English and Welsh trawling trade could boast a fleet of 1,132 vessels but the Admiralty purchased a large number some four weeks prior to the outbreak of war. Some 250 Grimsby trawlers alone were bought or requisitioned as war loomed—though a number of the older ones were to be returned to fishing after a few months.[1] In the week leading up to the declaration no less than 227 trawlers from English and Welsh ports—in addition to many Scottish vessels—were taken over by the Admiralty and requisitioning continued at a high rate throughout the rest of the year. By the end of December about half the fleet was in Admiralty hands.[2]

The Admiralty's appetite for trawlers remained considerable throughout the war and altogether about 816 English and Welsh trawlers were requisitioned at one time or another. In addition, about 200 steam drifters were also taken into naval service but the maximum number of trawlers requisitioned at any one time was about 690.[3] Scotland contributed a further 188 steam trawlers and liners to the war effort or about fifty-five per cent of the available fleet. This was proportionately smaller than the English and Welsh figure largely because of the smaller size and greater age of the fleet.

Additionally, seventy-seven per cent of the Scottish steam drifter fleet was requisitioned along with 240 motor boats.[4] Admiralty demand varied throughout the conflict. Towards the end of 1939, when the magnetic mine first appeared, the need for minesweepers increased, causing an intensified spate of requisitioning, but once counter-measures were devised a number of these vessels were returned to the fishing industry. However, the events surrounding the fall of France and its aftermath, and the steady loss of trawlers on war service, stretched naval resources to the limit; by early 1941 the Admiralty was making its maximum wartime demands on the fishing fleet. Thereafter a limited number were released to the trade once more.

The strength of English and Welsh trawler fleet actually available for fishing during the conflict was generally at about twenty-five per cent of pre-war levels, but in reality the catching power was much less, for the Admiralty deliberately requisitioned most of the larger and more efficient modern vessels. The fish trade was left with the more elderly vessels—a large number of which should, in any case, have been replaced ten or more years before the war had the ports pursued an effective replacement and modernization policy—and these tended to need lengthy periods out of service undergoing major repairs. Moreover much of the North Sea—except for a strip down the east coast which varied in width between fifteen and thirty miles and on the inside of which was a mine belt—was closed to shipping, as were the distant-water grounds off Bear Island and the Norwegian Coast; this led to a radical redistribution of the shrunken fleet.

Hull was hit harder than any other trawling centre (see Table 5). Having by far the most modern fleet at the outbreak of war, it suffered heavily through Admiralty requisitioning and many of its more elderly vessels were then transferred elsewhere.[6] From the end of 1940 it had virtually ceased to operate as a trawling centre. Such moves caused great hardship, especially in the early stages of the war before labour could be more effectively deployed. Each trawler supported more than just its crew of twenty for a large number of jobs in ancillary trades relied on the trawling fleet. A large swathe of the port's Hessle Road workforce was thus thrown out of work by the end of 1939 as fish merchants and processing houses were left without adequate supplies.

Table 5. Trawlers still Fishing—Various English Ports 1939–1944

	July 1939	*Dec. 1939*	*Dec. 1940*	*Dec. 1941*	*Dec. 1942*	*Dec. 1943*	*Dec. 1944*
North Shields	84	36	20	17	16	17	17
Hartlepool	—	6	3	2	2	2	2
Scarborough	7	7	5	5	4	3	4
Hull	191	66	1	7	18	18	31
Grimsby	381	183	91	75	66	74	90
Lowestoft	46	2	—	—	—	—	—
South East Ports							
Brixham	6	8	—	—	—	—	—
Plymouth	3	3	1	2	4	4	4
Cardiff	13	5	3	4	5	5	5
Swansea	17	14	22	16	18	15	11
Milford	109	51	64	41	41	41	40
Fleetwood	174	101	80	93	91	83	72
Total	1,031	482	290	262	265	262	276

Source: *Fisheries in Wartime.*

The gradual westward movement of vessels made Fleetwood the premier British trawling port for a while. The principal grounds fished during the war were off Iceland, the Irish Sea, the west coast of Scotland and the Hebrides, as well as north-west Ireland together with the prolific hake grounds off that country's south-western coasts. With the exception of Iceland all were easily worked from Milford Haven or Fleetwood and the west-coast ports made a proportionally larger contribution to the country's fish supply than ever before.[7]

Fishing operations were also affected by enemy action; it was soon evident the Germans were again regarding fishing boats as legitimate targets. In the early months of the conflict a number of trawlers, particularly those working off the north and north-west coasts of Ireland, were sunk by gunfire from U-boats. On 15 September 1939, for instance, the Fleetwood trawler *Davarra* was working off Tory Island in daylight when a U-boat surfaced and began firing. The commander stopped once it became obvious that the trawler's crew were abandoning ship and taking to their open boat, but afterwards spent half an hour putting up to thirty shells into the stricken trawler before slipping beneath the waves.[8]

The principal U-boat phase of the fishermen's war lasted about three and a half months but towards the end of 1939 aircraft attacks increased in regularity. On 12 December the Grimsby trawler *Pearl* was overwhelmed by three German planes whilst fishing sixty miles from the Humber. The crew found it impossible to abandon ship as the Germans continually strafed it with bombs and bullets and only escaped from the rapidly sinking trawler when the attackers made off. During the same morning two Granton trawlers, *Colleague* and *Compagnus* were attacked by Dorniers and the *Compagnus* was sunk. Later in the day another Granton vessel, *Isabella Craig*, was attacked in a similar fashion and her crew reported being machine gunned after taking to their open boat. A further Granton vessel, *Trinity NB* suffered the same fate a day or two later and her fireman was drowned after the crew were forced to leap into a wintery sea.[9]

After the fall of France, trawlers working off the southern coasts of Ireland were also subjected to air attacks which continued during most of 1941. But throughout the war mines remained the most deadly enemy. Sea mines sank more fishing vessels than any other weapon. Out of the eighty-five English and Welsh steam trawlers lost to enemy action whilst fishing, thirteen were accounted for by U-boat gunfire, twenty-three by air attack and twenty-five by mines. A further five steam drifters and twenty motor fishing vessels were also sunk by the enemy. Eighty-eight skippers and 739 mates and men—827 in all—were lost with them. About 200 attacks were also made on Scottish fishing vessels and twenty-nine were sunk; at least 174 working fishermen from north of the border are known or presumed to have lost their lives through enemy action whilst working.

The bulk of the trawlers sunk whilst fishing were lost in the first two years of the war, but from the beginning measures were taken to protect the fleet. In September 1939 a programme of forming fishing fleets into sections of from four to eight vessels, of which two were armed with twelve-pound guns, was introduced. The strategy was adopted at Hull, Grimsby and, to a lesser extent, at Fleetwood but was rejected, chiefly by the fishermen themselves, at North Shields, Swansea and Milford Haven, some of whom felt they were more likely to be attacked if armed. In fact no vessel sailing in these armed sections was lost to submarine or aircraft attack over the winter of 1939/40 but in May 1940 those trawlers which had been fitted with guns were swiftly requisitioned and sent to help with the evacuation of Dunkirk and elsewhere.

After the twelve-pounder trawlers were taken, various armaments were fitted including Lewis guns and other types of automatic weapons as well as kites, rockets and finally Oerlikon guns as they became available. Steps were also taken to neutralize vessels against magnetic mines whilst wheelhouses were reinforced, often with concrete, and fishermen gunners were given steel hats and weapons training.[10] Vessels were instructed to fish in company wherever possible but faced heavy attacks from enemy aircraft over the following year. As fishermen gunners became more proficient and armament was increased, many of the raiders were successfully repulsed. There were several grim battles between aircraft and trawlers from which the enemy came off second best. The ST *River Ytham*, for example, skippered by F. T. R. Kirby of Hull but working out of Fleetwood, was attacked whilst fishing off Faroe in July 1941. A bomb narrowly missed the trawler, shaking it from stem to stern, and aircraft then raked the vessel with bullets from bridge aft to galley. A steam pipe was fractured and several of the crew sustained gun and schrapnel injuries, including Skipper Kirby who collected two leg wounds. Despite this he took over the machine gun from the dazed gunner and, screened by the escaping steam, emptied a full pan of ammunition into the attacking aircraft which then veered off belching black smoke and crashed into the sea.[11]

Fishermen also saved many lives at sea, rescuing crews from stricken ships as well as British and enemy aircraft. Between 1941 and 1944 English and Welsh vessels rescued no less than 389 airmen[12] whilst Scottish vessels, such as the ST *Swallow* which saved fourteen men from a Sunderland flying boat, also brought many safely to shore.[13] On 26 March 1941 the Grimsby trawler *Pelican*, running through thick weather and a heavy swell close to the Longstone Light on the Farne Islands, came across the liner SS *Somali* in dire straits. The *Somali* was ablaze and the fire was out of control. A destroyer standing by had been badly damaged attempting a rescue and asked the *Pelican* to try and take off the crew. Skipper R. Carrick managed to keep his trawler alongside the stricken liner for forty minutes despite the heavy seas. He later ran aground but was able to get off and limp safely into the Tyne.[14] Skipper Carrick was also awarded the MBE. By the end of 1946 some seventy-three fishermen who were not on naval service had been awarded the MBE whilst another ninety had gained the BEM for actions during the war.

In 1914 many fishermen, caught up in the surge of patriotism which accompanied the outbreak of war, had joined the army; their vocational expertise was consequently lost to the country at a time when there was an acute shortage of seafarers to maintain food supplies and keep the shipping lanes clear of mines. Others, who had signed up for the Royal Navy, found themselves on large ships when they were better suited to working on smaller vessels. In the late 1930s much thought was given to the effective deployment of fishermen in any future conflict and when the first Schedule of Reserved Occupations was drawn up all classes of fishermen were reserved from the age of eighteen years, except for service in the navy. Though quite a number joined the Merchant Navy or the Royal Navy of their own accord, the policies pursued throughout much of the Second World War meant that most fishermen continued fishing until required for naval service, and when recruited, were usually used on small craft—particularly those of the Patrol Service—where their particular skills could be best deployed.

As vessels were requisitioned the numbers of men left fishing also shrank but by no means at the same rate. The large numbers of trawlers taken by the Admiralty during the summer and autumn of 1939 far outstripped the recruitment of fishermen—even though large numbers were called up as naval reservists or volunteered for service—and fishermen in some English trawling ports found themselves facing a period of unemployment. This situation had changed dramatically by 1941 and, henceforward, personnel were at a premium. In Scotland over 8,000 men from all branches of the fishery—or nearly fifty per cent of the total fishing force in 1939—were on armed service by 1942 whilst another 2,000 were employed for the duration in the Merchant Navy or other war-related occupations.[15] Figures for the war years are not available for England and Wales but it seems likely that an even greater proportion of the labour force was recruited for war service. Their places at sea were increasingly taken by boys and elderly men and those left of military age included a large proportion not classed as A1 fit. Several trawlers were reported to be fishing with skippers in their seventies and crewmen over eighty were by no means unknown. Later in the war, as a number of vessels were released for fishing, steps were also taken to release fishermen from naval service to crew them.

The war had a profound effect on the British fish supply. The total supply from all sources, fresh or frozen, amounted to 22,417,780 cwt in 1938 but by 1941 this had slumped to 7,771,016 or thirty-five per cent of the 1938 total. Thereafter, there was a gradual improvement and by 1944 supplies were running at forty-eight per cent of the pre-war total. The make up and sources of war-time fish supplies differed markedly from the preceding years of peace. British landings of demersal fish—caught principally by steam trawlers—slumped more dramatically. In 1938 they had amounted to 15,148,145 cwt but by 1941 they had fallen to 3,640,669 cwt or twenty-four per cent of pre-war levels. By 1944 they had risen 4,744,644 cwt but this was still only thirty per cent of 1938 figures. Nor was the overall shortfall between total and British demersal landings made up by landings of pelagic fish—herring and the like—for by 1941 they had slumped to twenty-four per cent of pre-war landings and only risen to thirty per cent by 1944. The balance was made up by increased imports of demersal fish. These can be broadly split into direct landings by foreign fishing vessels and consignments arriving by cargo boat. Direct landings rose from 371,660 cwt in 1938 to 1,588,797 cwt in 1944, an increase in the region of 500 per cent, whilst cargo imports rose from 751,864 cwt to 2,317,505 cwt, or by 308 per cent.[16]

After the outbreak of war a considerable number of foreign fishing vessels found their way to Britain. In the early months of the conflict, Belgian motor and steam trawlers were landing at Fleetwood, Milford Haven, Brixham and other places whilst Danish motor seiners put into Hull and Grimsby. After Denmark was invaded in April 1940 the Danish fishing vessels were seized as prizes; Denmark under German occupation being classed technically as an enemy country. A number of these craft were eventually put to fishing under the British flag. After the invasion of Norway a considerable number of Norwegian fishing vessels escaped and many then worked out of Scottish ports, in particular Buckie. When the Germans marched through Holland and Belgium many more fishing vessels fled to Britain, often carrying refugees, and these were later joined by French boats. Up to two-thirds of these were requisitioned for naval purposes but the remainder were encouraged to fish under licence from south- and west-coast ports. Most of the 180 Belgian vessels were based at Brixham, Newlyn and Milford Haven though a few of the larger craft fished from Fleetwood, Cardiff and Swansea. About twenty Dutch steam trawlers were similarly allocated to Fleetwood.

However, the largest imports of fish—both direct landings and cargo—came from Iceland. In 1938 Icelandic imports comprised just twenty-seven per cent of all British demersal fish imports but by 1944 they made up seventy-one per cent of the far larger total.[17] This trade was not conducted without danger and vessels engaged in the trade suffered a number of serious attacks, but it proved of great benefit to the development of the Icelandic fishing industry. Icelandic imports were made principally by steam trawlers, which preferred to land at the less exposed port of Fleetwood, and by carriers landing at Fleetwood, Hull and Grimsby.

Not surprisingly, given the shortfall in supplies and the rationing of many other foodstuffs, fish, which remained unrationed, rose very sharply in price through 1940 and continued to rise until price controls were introduced in mid-1941. By that time Scottish cod was at more than four times its pre-war price and saith—a fish previously held in little regard—was worth up to ten times its pre-war value.[18] Thereafter, there was a greater degree of stabilization in fish prices but controls were not relaxed until several years after the end of the war.

Over the war years the catching efficiency of the fishing fleet improved markedly despite the fact that the trade had been deprived of many of the younger, fitter men, and that those left behind had to fish under severe restrictions, on relatively obsolete craft, and were subject to enemy attack at almost any time. It became widespread practice to head the fish at sea and land only their bodies, which increased the quantity landed out of a given catch weight compared with years when unheaded fish were landed. It would appear that conservation through war encourages a marked increase in the weight of fish on the grounds, even though superficially, the trade may be operating in a less 'efficient' manner than in peacetime. The same effect had been noted as a result of the 1914–18 war; it is sad that what was learned from such lessons was not put into practice when peace returned.

The years 1940 and 1941 were probably the worst of the war for the British trawling trade. Working vessels were subject to intense enemy attack; indeed, nearly two-thirds of the English and Welsh trawlers lost through enemy action whilst fishing over the six years of war went down in 1940 and 1941. Admiralty demands, in terms of requisitioning vessels and conscripting crews, stripped the trade to the bone. Fishing on the North Sea grounds was particularly restricted and Hull, of course, was particularly badly affected. It is not surprising that catch levels during 1941 were the lowest of the war. Thereafter, matters slowly improved. Vessels were better able to defend themselves and Coastal Command aircraft provided valuable protection. From the end of 1941 losses from submarine and aircraft attack were drastically reduced. Catches edged upwards and during 1944 extra vessels and crews were being released to join the working fleets. Gradually, slowly, the tide turned in the fishermen's war but only peace could bring a return to normality. Yet whilst the war raged, trawlermen still had to contend with their old enemies, the elements: thirty-nine more steam trawlers were lost over these years to what might be classes as traditional causes.[19]

Most trawlers and trawlermen drawn into the forces were directed, of course, into the Royal Naval Patrol Service. The Patrol Service has sometimes been described as a navy within the Royal Navy 'right down to the exclusive silver badge worn by its seagoing officers and ratings'.[20] Patrol Service headquarters were in Sparrow's Nest, a seaside municipal pleasure gardens at Lowestoft, and in the early stages of the war newly recruited fishermen were joined as ratings by those who manned the tugs, barges and lighters of Britain's ports, harbours, rivers and coastal waters. Many of the officers were fishing skippers, although a number of Royal Navy and RNR personnel were included in senior ranks to maintain a degree of Senior Service discipline.

The Patrol Service started with 6,000 men and 600 vessels but eventually grew to 66,000 men and 6,000 vessels of all shapes and description. Trawlers converted into warships required up to double their fishing crew strength.[21] The ranks of seafarers were therefore gradually supplemented by landsmen from all walks of life, many of whom had never set foot on a boat before. This made for some interesting mixes of crew. They soon earned the soubriquet of 'Harry Tait's Navy' after a famous music-hall comedian who was always confused and confounded by modern devices.[22] The initial ranks of experienced seafarers—in particular the trawlermen, who came from a tradition far removed from that of the Senior Service—paved the way for those who followed, and those officers who tried to enforce rigid Royal Navy discipline sometimes found life almost impossible. But whilst the rest of the Royal Navy viewed this apparently ill-disciplined and haphazard fleet with what amounted at times to disbelief, its vessels and men sailed forth to face all that the enemy could throw at them with a level of courage, skill and determination that earned them the respect of all.

Royal Naval Patrol Service vessels played a crucial role as minesweepers and losses of vessels and men through this activity remained high throughout the war. Many were fitted with asdic, an early form of sonar, and worked on anti-submarine duties sending several U-boats to the bottom. They also saw action in many theatres of war and were used as escorts on the hazardous Russian convoys. Some fourteen trawlers were lost during the ill-fated Norwegian campaign in April 1940 during which Lieutenant Commander Stannard, an ex-merchant officer, won the VC when he put his Hull trawler *Arab* against the blazing pier at Namsos to fight a fire that at any moment threatened an ammunition dump. Another trawler, the *St Cathan*, sailed for Norway with just one four-inch gun, a Lewis Gun and 350 rounds of ammunition, survived no less than twenty attacks from enemy bombers, and arrived back at Londonderry with her crew armed to the teeth and decks covered in loot.[23] The trawlers involved were not only subject to unremitting attack but faced acute shortages of supplies and were forced to scavenge. The Grimsby trawler *King Sol* came up against an abandoned troopship—a former liner—in a fjord and was able to raid it for vital supplies. The crew also found a grand piano on board and, after straddling it across their open boat and sailing back to the *King Sol*, winched it aboard. A Torquay crew member organized its stripping down and reassembly below deck and for much of the war the vessel was able to offer visiting crews a higher level of entertainment. *King Sol* survived several enemy attacks during the campaign and also returned to Britain loaded with army equipment salvaged from war-ravaged quaysides by the crew.[24]

Fishing vessels, both civilian and Patrol Service, played a major role in the evacuations from Dunkirk and elsewhere on the French coast the following month. Working fishermen from all branches of the trade as far north as Whitby answered the call.[25] Some fishing vessels went across with their own crews and many vessels ran to the mole under heavy attack and picked up soldiers whom they either ferried straight back to England or transferred to larger vessels offshore. By the end of the Dunkirk evacuation at least fourteen Patrol Service trawlers and drifters had been lost. Two weeks later, during the evacuation of St Nazaire, the Grimsby trawler *Cambridgeshire*, skippered by Billy Euston, was the only vessel to reach the stricken troopship *Lancastria* as she went down with four thousand on board within twenty minutes of a German bomber raid. *Cambridgeshire* was able to save 1,009 lives and was described as seeming just a solid mass of survivors when she transferred them to an offshore transport. Later that night the trawler quietly steamed under the bombarding German guns and into St Nazaire harbour to pick up General Staff from the British Expeditionary Force.[26]

Trawlers from the Patrol Service saw action across the world. They worked with convoys on the Atlantic and Russian runs; five were lost when sent to the eastern seaboard of the United States to help with anti-submarine duties there in early 1942 and some later saw duty in South African waters. During the long seige of Malta only the trawler *Beryl* survived when all her sister ships were sunk by enemy bombers, and she was thereafter known by the islanders as the flagship of Malta. *Beryl* was the only warship to serve throughout the seige of Malta and was in commission in the Mediterranean longer than any other ship in the Navy.[27]

With the return of peace it was evident that the Royal Naval Patrol Service had played a key role in the victory at sea and had paid a high price in consequence. Although it is not possible to single out exactly how many fishermen died whilst serving with the Royal Navy it is known that some 2,385 officers and men of the Royal Naval Patrol Service, aged from sixteen to over sixty lost their lives.[28] The Patrol Service lost nearly 500 vessels including more than 400 trawlers, drifters and whalers. These losses were far greater than any other branch of the Royal Navy.

The fish trade as a whole and working fishermen in particular had also paid a heavy price. At least 1,243 British fishermen lost their lives whilst following their livelihood during the war.[29] The effect can be gauged a little more accurately at certain ports: Hull, for example, boasted a fleet of 191 trawlers in July 1939 but by VE day ninety-six or just over half had been lost. Grimsby lost over 600 fishermen whilst fishing or on naval service during the same period.[31] Few trades can be said to have given more.

14

The Cost of Trawling

The pace of the trade's post-war recovery was largely governed by the release of trawlers from naval service and the limited rate at which shipyards could turn out replacements for those lost in action. All ports gradually recovered. Though Hull had suffered most from Admiralty requisitioning and war losses it soon acquired several ex-Grimsby trawlers and converted a number of former Patrol Service vessels. Local shipyards began delivering new vessels within the year and by the end of 1946, 136 trawlers were working out of the port compared with about 191 at the outbreak of war.

Six years of conservation through conflict had given stocks on many grounds a chance to recover and as fishermen returned in growing numbers to their old haunts they were rewarded with bountiful catches. Post-war food shortages and rationing increased demand for unrationed fish and catches normally sold at close to the wartime maximum prices set and still enforced by the Ministry of Food.

There was plenty of money to be made from big landings and the distant-water fleet maximized output to satisfy heavy demand. The Bear Island and Barents Sea grounds proved especially prolific and trawls were often towed for only the shortest of time before being hauled aboard with cod ends filled to capacity. As one former trawlerman recalled:

> It was more or less a question of paying away, getting the warps in the block, knocking out and you'd got a full trawl. You were only on bottom about five minutes at most.[1]

The wartime process of heading fish at sea continued for some while so that trawlers maximized the amount of edible fish flesh they could cram in their fish-rooms. Distant-water trawlers found that big catches and large profits were there for the taking. In 1946, for example, skipper A. J. Lewis in *Saint Bartholomew* landed 4,225 kit—or over 264 tons—of fish which brought £17,302 on the market. Iceland was comparatively neglected during this post-war bonanza whilst Bear Island, the Barents Sea and Norwegian coast accounted for 82.2 per cent of the Hull catch in 1947.[2] Such prolific conditions did not last for ever and soon trawls were more usually on the bottom for upwards of an hour. Before 1950 two or three hours was becoming commonplace once more. Attention began to switch back to Iceland. Even so, the trawlers found these grounds well worth the long voyage for years to come.

The post-war boom merely disguised rather than displaced the white fish industry's deeper underlying problems. The near- and middle-water sectors were the first to experience the old troubles; North Sea catches began to show renewed signs of decline even before the end of 1946. Little was being done to tackle the questions of obsolete fleets and low long-term yields. As late as 1952 no less than 637 vessels out of the British near- and middle-water fleet of 817 vessels over 70 foot in length had been built before 1921. No more than fifty-six had entered service since January 1945 (see Table 6). The rate of replacement was so low that it would have taken forty more years to replace all vessels built before 1921.[3] Profit margins were generally so low in the middle and inshore sectors that there seemed little chance of their fleets being rebuilt if left to their own resources.

Fish prices collapsed in November 1949, a few months after the Government had removed price controls, and this brought unemployment and distress to the near- and middle-water sectors. The Government responded by finally creating the White Fish Authority which, together with its older counterpart the Herring Industry Board, became responsible for overseeing a programme of grants, loans and subsidies aimed at restoring the profitability of the near- and middle-water fleets. This strategy enabled a programme of modernization to get underway but by then absolescence was so widespread that as late as 1960 there were still 148 near- and middle-water vessels fishing which had been built before the end of the 1914–1918 war.[4]

Table 6. Number of Trawlers 70 to 140 feet in Length on 31 December 1951

	England & Wales	*Scotland*	*Total*
Built Before 1921	419	218	637
1921–1940	71	35	106
1941–1944	11	7	18
1945–1951	50	6	56
Total	551	266	817

Source: WFA Annual Report y/e 31 March 1952.

The collapse in fish prices was in part due to the growing volume of distant-water landings coupled with an increase in imports. At the same time demand started to falter when consumers turned away from fish, as food shortages and rationing, particularly of meat and eggs, eased. Though landings from the near- and middle-water sectors were in decline, total supplies of white fish coming onto the market in 1951 were twelve per cent higher than in 1938. In contrast, distant-water trawlers landed eleven per cent more fish in 1951 than they had done in 1938 whilst imports showed a 139 per cent rise over the same period, though they still accounted for only seventeen per cent of the market as late as 1949.[5] The formidably efficient distant-water sector was better placed to weather such storms. Its history of profitability was reflected in the age distribution of its fleet—still based primarily on Hull, Grimsby and Fleetwood—which contrasted markedly with that of the near- and middle-water sectors (see Table 7).

Table 7. Age Distribution of Distant-Water Fleet* on 31 December 1951

	England & Wales	*Scotland*	*Total*
Built Before 1921	2	—	2
1921–1930	58	—	58
1931–1940	119	—	119
1941–1946	27	2	29
1946–1951	82	2	84
Total	288	4	292

Source: WFA Annual Report y/e 31 March 1952.
*Vessels over 140 ft in length

Because the distant-water fleets 'paid their way' they were not included in the Government fishing industry subsidy scheme for the near- and middle-water fishing vessels. In the early 1950s over-supply was primarily a summer problem for the distant-water companies and—as in 1938—they attempted to reduce the seasonal surplus, and maintain prices, by co-operating in a scheme to lay up vessels during the summer months when supply problems were at their most acute.[6]

Co-operation was more easily achieved in the distant-water sector because ownership was far more concentrated than amongst the near- and middle-water fleets. At Hull, for example, which was the leading distant-water port, ownership of the entire fleet lay in the hands of fifteen to twenty companies during the later 1940s. The port's trawler owners had, of course, a long tradition of close collaboration with fish curers and wholesalers in the development and operation of ancillary concerns including ice factories, fish meal and cod-liver oil plants, as well as marine engineering facilities. They could also mobilize sufficient resources to maintain a programme of investment in new vessels. Although Grimsby and Fleetwood were often criticized for concentrating on smaller companies,[7] by the 1950s their deep-water sectors were mainly controlled by larger firms, unlike the near- and middle-water fisheries where small companies and single-owner vessels abounded.

The distant-water quest for high levels of efficiency was, however, pursued in a manner which placed great pressure not only on fish stocks but also on the men who fished them. Throughout the post-war period, until the trade's virtual demise, many trawlermen continued to live and work in conditions that few modern landsmen could or would tolerate.

Unlike the early smacks, distant-water steam trawling required a varied crew with a range of specialist skills. A typical post-war side trawler—or side winder as they were soon known—shipped a skipper, mate and bosun in addition to a chief and second engineer, two firemen, a radio operator, cook and galley boy. The deck crew consisted of a third hand, eight spare hands and a 'deckie learner'. Taken together, the crew formed a complex and, to the outsider at least, somewhat confusing social hierarchy.

A trawlerman's earnings depended greatly on the performance of his vessel. This in its turn was determined by a range of factors which affected the quality and quantity of catches and the state of market demand when landing. In the case of the skipper his income was principally—if not solely—determined by his share of the proceeds of each trip he commanded. At Grimsby in 1946 a skipper would receive 1⅜ of fourteen shares of his trawler's net profits—that is earnings after agreed expenses have been deducted; in other words £9 16s 0d in every £100 net. His counterpart at Hull received the same proportion plus one per cent trip money on the gross, whilst Fleetwood skippers were paid ten per cent of the net revenue of the vessel.[8] These shares were modified over the years and by 1959 Hull skippers took ten per cent of net profits, one per cent of the gross and a thirty-five shilling per week minimum wage.[9] The mate's wages were also largely determined on a share basis: a fourteenth share of net earnings being usual at both Hull and Grimsby in 1946. The rest of the crew were paid a mix of basic and poundage depending on their position in the crew's complex hierarchy. The chief engineer, for example, though classed as an officer, was ranked in terms of command below the third hand, and yet he generally earned just a little less than the bosun. The cook was paid at a similar rate to the deckhands whilst the firemen had their own grade of pay. In 1946 Hull and Grimsby deckhands received £4 12s 9d per week basic including War Risk Money and 2½d in the £ on net earnings.[10] In a later formula, War Risk payments were abolished and poundage payments were determined on a trawler's gross earnings. The share element made for great fluctuations in earnings.

In early February 1946, for example, at the height of the post-war boom, the skipper of the Hull trawler *Imperialist* earned £1,277 1s 2d for a twenty-two day trip whilst his mate took home £834 14s 3d and the spare hands £156 16s 3d, or an average of £49 18s 1d per week; astonishing earnings when the average manual wage was still well under £10 per week. Yet that same vessel, on returning from a twenty day trip the following June, made its skipper and mate only £50 12s 0d and £32 2s 0d respectively. On that occasion the spare hands settled with £25 4s 0d, the equivalent of £8 19s 0d per week.[11] A skipper therefore had the opportunity of huge earnings. Indeed, in the boom conditions prevailing just after the war it was reported that some skippers were making more than £5,000 a year and these continued to do well in the following decades as well; in the first week of August 1960, for example, five Hull skippers landed trips which grossed over £10,000, earning them at least £800 apiece.[12] On the other hand, a poor trip which barely covered expenses would leave a skipper with little if anything to take ashore and a short run of barren trips could see him out of a ship.

Trawler owners usually aimed to maintain efficiency and profits by operating all factors of production—labour as well as capital—at maximum levels. In the early 1950s a distant-water trawler would normally remain in port between trips for between forty-eight and sixty hours; by the end of the 1960s, after union pressure, the usual was still only between sixty and seventy-two hours.[13] In some years vessels might be laid up for part of the summer in reaction to low landing prices, but unless they required a major refit, suffered serious mechanical failure, or sustained damage at sea, most trawlers would spend just two further weeks in port undergoing an annual survey. The rest of their working life was spent on the grounds or voyaging to and from them.

Crews were expected to work with an almost equal rigour. The average trawlerman completed about thirteen trips a year, many on the same vessel. For much of the year he spent just two or three days at home while his vessel turned round before heading back to sea. Once at sea the estimates vary with regard to the hours actually worked. An estimate given to a 1946 Court of Inquiry claimed the maximum to be an average of 13.2 hours per day[14]—92.4 hours per week—over a trip uninterrupted by bad weather. At the end of the 1950s Tunstall similarly estimated that trawlermen worked a ninety-three hour week.[15] However, the owners argued that if fishing operations were suspended through bad weather then the hours worked dropped whilst the men spent their time, if not on watch, in their bunks.

It is difficult to determine a distant-water trawlerman's effort strictly in terms of hours worked per trip as he could not leave his working environment to enjoy his 'leisure' in the conventional sense. Moreover, working patterns for all aboard were far more complex and variable than was the case for most landsmen. During the post-war period the engine-room personnel worked eighty-four hours per week in six-hour watches throughout the voyage. Though most of their work, except at times of mechanical breakdown, was of a routine nature and—particularly after the demise of the coal-burning steam trawlers—less arduous than in pre-war days, it nevertheless remained far from pleasant; each watch being carried out below decks in a noisy, hot and often pitching engine room. The cook meantime might be on duty from 5.30 am in the morning to 8 pm at night preparing meals and refreshments in his wildly pitching galley.

A deckhand's voyage was far more variable. During the five or six days spent steaming off to the grounds the deck crew were divided into three watches and covered eight hours each day in two four-hour spells. Usually one full day, known as a 'field day', was devoted to repairing and preparing fishing gear for action. The old coal burners usually carried coal for the voyage out in the after fish-room which would have to be washed or swilled out and the pound boards fitted before the grounds were reached.

Once on the grounds trawling was soon under way and work rates increased in terms of both pace and physical effort. The deck team worked rapidly on their allotted tasks. The warp wires which towed the net ran from the winch round a system of sheaves to the gallows on the ship's side and left the vessel via blocks on the gallows before streaming astern roughly parallel with the vessel's side. During trawling the two warp wires were clipped together outboard of the gallows and held towards the ship's side by a type of snatch towing-block carried on the end of a short length of chain. When the crew prepared to haul in the trawl the block had to be 'knocked out' or released with a blow from a hammer. The job of securing at the beginning of tow and knocking out before hauling called for a deckhand to lean well outboard of the gunwhale of an often vigorously rolling ship. When the trawling gear was being hauled in some deckhands operated the winch dragging up the gear whilst others made ready to secure the massive steel-bound otter boards to the gallows as they came up the side. Once the otter boards reached the gallows head a deckhand had to climb up, secure the board to its chain and hook, then unshackle the warp. The net was then pulled on board, usually over the starboard side, followed by the large bobbins surrounding the mouth of the net. The cod end was dragged close to the vessel side and winched aboard before the mate loosened the restraining knot letting the fish spill on to the deck. Such operations, involving the securing of cumber-

Plate 7. The Grimsby Steam Trawler Northern Duke

some equipment amidst moving wires often under considerable tension and by open winches, were often conducted at night on an open and pitching deck exposed to all the northern elements.

The entire operation was conducted as swiftly as possible but the time which elapsed before the gear was heaved over the side again might vary from twenty minutes to several hours, depending on weather conditions, the size of the catch and how long it took to repair any damage to the net. Everyone was expected to work with great energy and efficiency. Anybody who was too slow or not considered to be pulling their weight could find themselves out of a ship at the end of the trip.

Once the gear was back in the water the deckhands turned to dealing with the catch from the last haul. These were gutted—and in the case of cod and haddock the livers were cut out for boiling—before the fish were thrown into the washer and passed down the chute into the fish-room where they were stowed away by three men led by the mate who was responsible for keeping the catch in good condition. In good weather an average catch might be dealt with in about one and a half hours and this could leave the men with as much as an hour to go aft for tea or a smoke or to snatch a rest in their bunks. With moderate hauls on a calm sunny day in the short Arctic summer, trawling might not seem too arduous a job, but the weather rarely stayed fine for long and the seas were usually choppy or rough. Moreover, in the long Arctic winter trawling was carried through almost twenty-four hours of darkness in bitingly cold and rough seas which constantly covered both trawler and deckhands gutting half-frozen fish with a white coating of ice.

Once on the grounds fishing continued, if possible, twenty-four hours a day for up to ten or eleven days. During this time the deck crew were divided into four watches, three of which would be on duty at any one time. This meant each deckhand was expected to be on duty for eighteen hours a day with just six hours off. Though this system had been common for years the right to six hours off was only conceded by the Hull owners to the union in late 1947 after a strike,[16] a Government inquiry and adverse publicity in the local press. A few skippers were alleged to be working crews until they virtually dropped. Unlike a workman on land, a trawlerman could not walk off the job. The skipper's commands were law and to disobey him at sea was a serious offence. In the nineteenth century, courts had been used to enforce the apprenticeship system and after the Second World War a constant stream of fishermen were still coming before magistrates for disobeying orders at sea. In October 1947 matters seem to have come to a head when sixteen hands of the trawler *Lord Gort* were taken before the court for refusing to shoot the trawl after nineteen hours of continuous fishing when the skipper failed to grant their request for a period of rest. The incident had occurred towards the end of a long trip and when brought before the court they were each fined £1 with a guinea costs.[17]

Even after a more hard-and-fast agreement on the eighteen hours on/six hours off issue had been secured the men could still be called to work longer hours in exceptional or unforeseen circumstances. In very severe conditions, when ice formed on a trawler's masts, rigging and superstructure, men had to be called out to chip it off with axes or hot-water hoses as the very stability of the vessel could be threatened.

Once fishing finished and the gear was stowed away for the voyage home the work rate slackened. Although fishermen were not expected to 'chip and paint' like merchant seamen they still had to stand the regular watches, and the ship had to be cleaned through from stem to stern.

It is true that fishing operations were often disrupted by the need to steam to new grounds and by bad weather which obliged the trawler to run for shelter or else dodge into the gale. Yet, it cannot be argued that a trawlerman spent much time in relaxation when wedged in his bunk bracing himself against the violent lurch of a ship battling through heavy seas in a force ten gale. It has sometimes been said that in the way of compensation, a trawlerman was provided with free board and lodging as well as a small liquor and cigarette allowance. Yet for most people a narrow bunk within one's place of work was hardly preferable to the comforts of home on a long-term basis. In any case a trawlerman had to buy his own mattress and bedclothes in addition to his protective clothing.

Therefore, though a trawlermen working throughout the year seems to have been well paid in terms of the amount of money he earned per year compared with the average post-war manual worker, it is clear that he had to put in far more hours in a much harsher environment. By any calculation his hourly rate of pay was extremely poor, being around five shillings an hour at the end of the 1950s. As Tunstall said in 1962:

> the fisherman's annual earnings are extremely high, but his basic hourly rate is low—probably lower than that of almost all adult workers in Britain today.[18]

This situation changed little throughout the rest of the 1960s. In 1968 it was estimated that deckhands on Hull distant-water side trawlers earned about £34 per seven-day-trip week or about 7s 3d hour for a ninety-three hour week,[19] far below the hourly rate of the average manual worker.

Even the crew's life ashore was usually governed by trawling's relentless timetable. If they intended signing on for the next trip then family and social life had to be fitted into the time it took to unload the trawler and make her ready for sea.

Once the vessel tied up on return to port the skipper handed over to the mate who supervised the unloading and liaised with the owners about servicing requirements and necessary repairs. At Hull the catch was landed by eight-man gangs called bobbers—at Grimsby they were known as lumpers. Work started at 2 am and usually all catches were unloaded in time for the auction which began at 8 am. The work was carried out in the often bitterly cold early hours because the catch would deteriorate more swiftly when temperatures rose after day-break. Four bobbers worked in the hold filling wicker baskets with five stones of fish. A fifth man operated a portable electric winch to haul up the baskets to another who swung them ashore where a seventh caught them and tipped them into aluminium tubs or kits holding ten stone. These kits were then hauled away to the auction. The bobbers unloaded fifty to sixty kits an hour and dumped the remaining fish-room ice into the dock. Some sorting out of fish in relation to species, size and age was carried out during landing, and before the auction commenced an inspection was carried out. Any deemed unfit for human consumption was despatched for the fish-meal works.[20] In Hull, Dutch auctions were—and indeed still are—carried out. Under this system the salesman starts at a high price and works down until the lot is bought by a merchant.

In the meantime, preparations for the next trip were already underway. The cod-liver oil was drawn off, the fish room washed out whilst its pound boards were removed and scrubbed. In St Andrews Dock, Hull, all these activities were carried out in the so-called wet side before the trawler was towed over to the dry side in order to be reprovisioned and made ready for sailing.

Once in dock the crew were usually keen to make the most of their limited time ashore and after a good trip money was not a problem. Though not all distant-water trawlermen drank, for many their time ashore revolved around pub and club. Tides usually governed their period of freedom ashore and, until the union gained an extension, those landing in Hull on the evening tide could expect up to sixty hours on land whilst morning arrivals might get just forty-eight. With time so short, many took taxis rather than go on foot. Evening arrivals at Grimsby might head for the Freemason's Arms, otherwise known as Cottie's. Those landing in Hull often visited one of the Hessle Road pubs like the Star and Garter—better known as Rainer's—the Halfway, or Criterion. Others would head home, get washed, changed and head straight out again if the pubs were still open. In Grimsby many trawlermen still lived in the close-knit terraces around the Freeman Street, Park Street, and Isaacs Hill areas until the communities were rehoused in high-rise flats and new housing estates.[21] Before the 1930s most Hull trawlermen lived in the Hessle Road district but by the fifties and sixties an increasing number lived further afield, often in surrounding housing estates or neighbouring villages such as Hessle.

On the first morning ashore any Hull trawlerman wishing to collect his free 'fry' of fish would be on the dock by about 7 am before the auctions started; otherwise it was necessary to be there at 11 am to settle, that is receive their share of the poundage. For many this might again be followed by a few pints in Hessle Road pubs until 3 pm, followed by a session in a local club until 5 pm. A short trip home might be followed by another session in the pub and the evening might finish off with a party back at someone's house. Many younger fishermen made the most of their short time ashore and the high-spirited intensity of their brief forays contrasted markedly with the monotony of the long sea trip. Trawlermen were often regarded as being profligate with money and free with 'backhanders' as they were known in Hull. They invariably drank in large rounds, treating friends and hangers-on alike, and it was by no means uncommon for a trawlerman to take a taxi around with him, not only to maximize drinking time when moving from pub to pub but also to make sure he got home; many taxi-drivers had to help the trawlermen home and up to bed.

In fact the backhander was almost an essential part of trawling life. Everyone knew that a good trip and plenty of money might be followed by a poor settlement next trip or a period 'walking about' ashore. Being generous to friends had advantages for all: the favour could be reciprocated when one's own luck was down. It was an example of a close camaraderie which set them apart from the rest of the town, as so often did their dress and language. Young trawlermen in particular had a style of their own: fishermen's suits certainly set them apart. In the 1960s Hull suits generally had twenty-two to twenty-four inch bottoms, straight legs and pleated jackets. They were usually powder blue and often had collars lined in black velvet. Grimsby suits were somewhat similar in outline but had their own distinctive, detailed style and trawlermen there sported a far wider range of colours including grey, pillar-box red, royal or sky blue. Such suits made them stand out—as did their language. Working-class waterfront districts have long been noted for their free use of bad language, but the fish dock vernacular made use of the trawlermen's life, experience and attitudes. Sex in the missionary position was known as 'having it off Board of Trade style'; a particular room in a city pub in Hull where women of reputedly questionable morality met was known as the 'Skate Pound', apparently because of a suggested similarity between the vagina and the mouth of a skate.

The second day ashore might follow a similar agenda of pub and club, then a last night in bed before speeding down to the dockside—often by the ubiquitous taxi—to catch the morning tide. In early post-war days when the shortage of beer threatened to curtail such activities, Hull trawlermen relied on the Hessle Road 'bush telegraph' for information on when and which pub had beer for sale. A taxi ride out to the Newbald or South Cave pubs with a few spare fries of fish might also mean fresh eggs and bacon to take back for the family, as well as a guaranteed supply of beer. It is true that many Hull trawlermen led a much more sober and relaxed life ashore but the variations on the theme outlined above were by no means uncommon.

This manner of living could wreak havoc with family life. Most trawlermen married early, had youngsters and made generous provision for their families. But their wives bore the brunt of raising the children. Running the family home was usually the wife's chief occupation as many trawlermen were opposed to them working. Moreover, though wives often shared in the hectic high life of their husband's brief time ashore, their virtual single-parent status tended to tie them to the house at other times. In many households the wife appeared to struggle alone with the day-to-day drudgery and problems of raising children whilst her husband intermittently returned only to spoil them with gifts and money as compensation for his absence. Her neighbours or relations often proved a valuable emotional prop, particularly in the tight-knit terraces of working-class areas off Hessle Road, Hull, or Freeman Street, Grimsby. Yet even a trawlerman's prolonged stay at home between trips might produce friction in a household accustomed to having just one parent at home. Many trawlermen enjoyed long and happy marriages but divorce rates were often well above average. Others eventually gave up the sea because their wives became ill or finally persuaded them to take a job ashore.

Despite the amount of time he spent at sea, a trawlerman was still classified as a casual worker, signing on for just one trip at a time. A century earlier the trade had been scouring the workhouses of England for young recruits who could be indentured to trawling for up to seven years. By the second half of the twentieth century trawlermen came mainly from the locality. It was widely believed that most people who took up distant-water trawling came from fishing families but Tunstall has shown, in the case of Hull at least, that the true family tradition of trawling was largely a myth. At the end of the 1950s his estimates suggest that no more than about a quarter of those who were employed as fishermen in the 1930s had sons who were fishermen at the end of the 1950s. Moreover, in an occupation with a true father-to-son tradition such as printing and until recently dockwork, fathers positively encourage their sons to follow them.[22] Many distant-water trawlermen did everything possibly to discourage their children from following them to sea.

Though there were undoubtedly many families that could trace trawling through at least three generations,[23] it is probably more accurate to say that most distant-water trawlermen came from families which worked in the wide range of maritime, transport and ancillary trades found in the commercial ports. Whilst one or two male members of a typical large family may have taken up trawling, their brothers, uncles and fathers might be employed in fish-processing factories, on the railways and docks or in the extensive marine engineering trades. Others might sail on 'big boats', as the merchant navy was sometimes known to fishermen. Though many such families discouraged their offspring from going to sea, youngsters were often attracted to the trade—probably because of the images of high wages, manliness and virility associated with it. Trawling was seen as a hard life for hard men. Most post-war trawlermen started quite young, often at sixteen or seventeen, and might typically work first as a barrow boy on the dock before signing on as a 'deckie learner'; others may have been attracted to the trade after taking a pleasure trip during their school holidays.

Once established in the trade, fishermen moved freely from one trawler to another: a survey of Hull ratings in 1967 showed that only twenty-three per cent stayed on the same vessel throughout the year.[24] There were always men looking for a ship after taking a few weeks off, whilst others might have been fired or have left voluntarily in search of better earnings. Either way, there were a number of men on the dock-side looking for berths, and trawlers needing replacement crew. Each trawler company still employed ship's runners to find and sign up crews.

To many, recruitment was one of the most devisive aspects of the whole job. High-earning vessels tended to retain stable crews and the mate and runner often worked together under the skipper's instructions when trying to fill a vacant berth. In other instances the recruitment was left almost entirely to the runner. This power to hire virtually at will placed him in a strong position when dealing with unemployed trawlermen, who were often desperate for a good berth, and this sometimes led to accusations of bribery. Many fishermen claimed that placing a fiver in the service book when handing it over to certain runners was a way of securing a berth; many claimed to have seen this done, but few admitted to having done it themselves. The truth behind such widespread assertions is difficult to gauge, especially in a world where loans and backhanders were an integral part of daily life. Indeed, though many runners were undoubtedly honest men—no case of corruption was ever proved against a runner—the whole system of dockside recruitment and dealings encouraged an air of distrust.

For much of twentieth-century British industry, trade unions became the principal vehicle by which the workforce sought to protect and improve its rates of pay and conditions. However, organizing deep-sea trawlermen had always been difficult, and many of the problems which had confronted trade unionism in the late nineteenth century were still apparent in the third quarter of the twentieth.

After the 1926 General Strike the TGWU, rather than the National Union of Seamen, became the dominant distant-water-trawling trade union in many ports. Though it is probable that the proportion of trawlermen who remained members of a union increased after the 1935 Hull Liver Oil Strike they remained difficult to organize effectively. Union membership still tended to show a dramatic rise when a dispute flared up, as was the case with the Hull Strike of 1946, and interest would fall away afterwards. Officials still found it difficult to collect subscriptions from a workforce scattered across the sea whose movements ashore were hard to account for. In any case many fishermen were more concerned during their brief forays ashore with living for today rather than planning for tomorrow. Those with ambition tended to aim at becoming a skipper rather than collectively try to better conditions through union activity. It is perhaps worth noting that the engineers, for whom the path of promotion to skipper was effectively blocked, were the most conscientious trade unionists. Many trawler owners were slow to tolerate trade unions and continually adopted a hard line.

Trawler owners were often widely disliked; many seemed to desire almost total control over the workforce. In many cases their 'abrasive and autocratic style'[25] bred widespread hatred amongst the workforce. With a casual workforce it was comparatively easy to enforce obedience. The tradition of blackballing a recalcitrant fishermen was reputed to be still widespread after the war. Many former trawlermen will tell of their passionate hatred for the owners. Nevertheless such feelings did not encourage the growth of union support that had become, for example, a feature of the coal mining communities. In the fishing industry there was a greater tendency to accept hard treatment, like the harsh conditions, as part of the job and this may even have contributed to its image as a hard man's occupation. The situation was little changed at the end of the 1960s when the Holland-Martin Report on trawler safety bewailed the industry's failure to establish a close relationship with its workforce.[26] It is perhaps not surprising, but quite tragic, that unions were generally less effective here than they were in many other areas of the post-war British economy.

Strikes did occur but they tended to be spontaneous and from the grass roots. The 1946 strike occurred mainly because of a temporary collapse in market prices which cut earnings. However, it focused discontent on related issues and after a subsequent Court of Inquiry the TGWU not only established agreement on the eighteen hours on/six hours off rule with the owners but also secured an increase in the basic wage whilst the basis of the crew's poundage payments were in future calculated from gross rather than net profit which crews saw as a fairer arrangement.[27] But, as in other times, once the immediate cause of the strike was over, interest in union activity dissipated and the chance of building on these gains was largely lost.

In the sailing smack era many top trawler owners had begun their lives as humble apprentices. In a post-war, highly capitalized distant-water trawling trade, dominated by a comparatively small group of distant-water, limited liability steam fishing companies, few entrants to the trade had any chance of becoming owners. Still, trawling remained one of those rare pursuits where to earn the highest reward one always started at the bottom. Top skippers could expect to earn large sums of money—as we have seen, a number were making £5,000 a year in 1946 whilst some of the very best were earning over £800 a trip in the late 1950s. Virtually everyone of them had gone to sea as a deckie learner.

In the absence of an apprenticeship system, skills and experience were accredited by gaining a 'ticket'. A Hull deckie learner needed to be over the age of sixteen, have put in more than 100 days at sea and attended a four-week course at Nautical College to obtain his deckhand's ticket. Many remained deckhands for the duration of their seagoing careers but others took further tickets to move up the ladder. Once a fisherman had reached twenty-one and had been four years at sea he could go on a college course and sit for his bosun's or even mate's ticket. Those who went straight for their mate's ticket still had to sail for some time as a bosun. A small number of mates, picked out by skippers and owners as suitable material, went on to take their skipper's ticket. Most had their skipper's ticket before reaching the age of thirty.

A skipper could thus reap rich rewards but to do so he had to reach and stay at the top of a highly competitive career structure. There were always more skippers than vessels and trawler owners were not noted for sentiment. Most company fleets contained a range of vessels of varying ages and levels of efficiency. The newer, faster vessels generally caught the most fish and made the best profits. They attracted the best crews and were commanded by the top skippers. At the other end of the fleet the older vessels, nearing the end of their working lives, usually did less well and had a higher turnover of skippers and crew. Newly qualified skippers might get their chance to command one of the older trawlers whilst serving as mate on other trips. Early promise might help them to consolidate their position and continued good fortune could lead to command of a more modern and efficient vessel thus improving their chances of further success. In contrast, a skipper who made a series of poor landings would often be demoted to an older and less efficient vessel or end up back on the deck as mate.

Top skippers had a nose for fish: their instinct and fund of fishing knowledge helped them to know when to change grounds and how to locate fish when others fared poorly. They knew that the only time a trawler was making money was when the net was over the side. Yet engine trouble or crew injury might force a vessel into port and play havoc with the best of plans. Then again, if they trawled on a ground which badly damaged the gear, a great deal of fishing time could be lost whilst laid up repairing the net, or the cost of its replacement would cut into the vessel's gross profits. Bad weather conditions could also disrupt fishing and trawlers often had to run for safety; but for a skipper out to make his name or under pressure after a series of marginal performances there was always a temptation to keep working. As one ex-skipper put it:

In your own mind you'd think I shouldn't be fishing. I know

> someone's going to get hurt. I know I shouldn't be shooting. But there in the back of your mind you'd know somebody else will . . . When you got home and the gaffers read your log and it says I was laid or dodging for twenty-four or thirty-six hours they'd ask you why when someone else was fishing. But he might have been three or four hundred miles away . . . There's all sort of pressures on you from leaving the dock to getting back.[28]

A skipper had also had to judge when to turn for home and what market to go for. It might be worth forgoing the opportunity of taking a few hundred extra kit in order to land on a market when few other trawlers were coming in. Likewise, one eye had to be kept on the forecast as bad weather might slow the return voyage and add to expenses. Throughout the trip a skipper constantly relied on his judgement and experience to make the right decisions. If those decisions regularly made good money for the owners then he was more likely to be secure. If not then his position was in jeopardy.

The surplus of skippers over ships meant that the owners had the upper hand in dealings with skippers. They decided which trawler the skipper would command and when and where he would fish. A skipper who argued or refused to adjust to their plans might be rewarded by a period 'walking about' on shore. Only the really top or 'Don' skippers with a long record of big landings which few could emulate could afford to be really independent and make their own minds up when or where to sail, and when to have a few trips off.

The extreme pressures of this hazardous occupation also took their toll in terms of health and safety. Like the old smackmen, trawlermen often suffered from a variety of occupational skin diseases, the most common of which were infections of the hands and salt-water boils on the wrists, generally caused by the skin chafing on the sleeves of oilskins hardened by salt water and slime. Though not classed as occupational diseases, fishermen also had a higher than average mortality rate from cancer of the lung and stomach, hypertension and bronchitis and suffered unduly from gastrointestinal illnesses. Though the death rate had dropped since the era of the trawling smack in the nineteenth century, trawling remained Britain's most dangerous occupation. In 1969 the Holland-Martin Report calculated that the Standard Mortality Rate for trawlermen of all ages was seventeen times that of the male population as a whole.[29] Fishermen in the 15–44 age group were even more at risk, being twenty times more likely to die as the result of an accident at work than was the average for all occupied men. Though some men remained at sea into their sixties it was predominantly a younger man's life with many coming ashore well before the age of fifty.

Between 1958 and 1967 alone no less than 208 men lost their lives on deep-sea trawlers. The figures fluctuated considerably year by year for, as in previous decades, the industry continued to be afflicted periodically by disasters involving the complete loss of vessels, often with all hands. In the twelve months ending with the loss of the *Laforey* and her entire crew on the Norwegian coast in February 1954, Grimsby lost five vessels and fifty-five men. Between 1940 and 1976 Hull lost some forty-two trawlers. Disasters such as the loss of the forty men with the *Lorella* and *Roderigo* off Iceland in 1954 hit the national headlines and brought home to the public something of the price of fish. But national interest was often all too fleeting.

The media were back on the fish docks in full strength early in 1968 when Hull was hit by a triple tragedy. Three trawlers and fifty-seven men were lost in the space of four weeks. The *St Romanus* was the first to go. She sailed from Hull on 10 January and fears for her safety grew when it was revealed that nothing had been heard of her for over a week, though a search was not ordered until a fortnight after she had left port. In all probability she was lost in the North Sea during a force ten gale on her second day out.

Less than two weeks later the veteran *Kingston Peridot* disappeared in appalling conditions off Iceland, her last message being broadcast on 26 January, exactly thirteen years to the day after the loss of the *Lorella* and *Roderigo*. On 4 February, the *Ross Cleveland* capsized in atrocious conditions whilst dodging into an icy force eleven gale in Isafjord. Other vessels saw her go and heard her skipper's last broadcast but were unable to go to her assistance in such weather. Just one man survived. The mate, Harry Eddom, on deck chipping ice, was pitched overboard dressed in full fisherman's oilskins when she capsized. He was pulled into an inflatable canopied life-raft launched hastily by two men who had rushed up from below clad only in their underclothes. There were no survival suits in the life-raft and when it eventually drifted ashore in Isafjord only the frost-bitten mate was left alive.

The loss of fifty-seven men in such circumstances in less than four weeks provoked a national outcry amongst press, public and parliamentarians alike. The trawler owners came in for heavy criticism over their apparently ambivalent attitude to health and safety. In Hull tempers ran high, but typically it was not the stubbornly reticent trawlermen or their union who headed the protest: a group of wives, led by Lilian Bilocca, gathered by St Andrew's Dock lock pit in an attempt to stop trawlers sailing for the Arctic. Restrained by policemen as the vessels slipped into the Humber, their desperate shouts brought them national celebrity overnight and a temporary press and television platform from which to berate the owners and argue the trawlermen's cause in blunt Hessle Road fashion. The subsequent Holland-Martin Inquiry recommended a batch of safeguards including a ban on trawling off Iceland's North Cape in winter, the provision of a mother ship and daily radio contact during the worst months of the year.

Though public concern centred on losses of ships with all hands, more men were probably killed in the post-war period by a steady sequence of isolated accidents which went on year after year and decade after decade. Given that on side-winders the deckhands worked on an open deck only a few feet above the sea in all sorts of weathers, it is perhaps not surprising that loss overboard was the commonest cause of death. Nor was it surprising that non-fatal accidents on trawlers were well above the industrial norm with slips, falls and entanglements with machinery being the commonest causes of injury. Fatigue caused by long hours of hard labour and little sleep made trawlermen more accident prone. A larger crew would have eased the situation but the owners argued this would be an uneconomic way to operate side-winders even though the TGWU pointed out that Icelandic trawlers were carrying thirty-man crews.[30] From a health and safety perspective the eventual way forward pointed in the direction of the stern trawler design where most men worked under cover, and safety records were far better. However, the distant-water fleet had only fifty-five large stern trawlers by 1969 and side-winders formed the backbone of the fleet in terms of vessel numbers until the trade's virtual demise after the mid-1970s.

Yet whatever precautions and improvements could have been effected, the extreme nature of distant-water trawling—like all fishing—would always carry with it a far greater element of hazard than most other modern occupations. Even today, few members of the general public really understand the true cost of fish to the fisherman.

15

Freezers and Factory Fleets

The 1950s and 1960s were a time of profound change for the world's fishing industries. Many traditional fishing nations greatly expanded their catching capacity whilst a number of other countries, with previously only a limited interest in fishing, were to develop large-scale, capital intensive fishing industries of their own. The nature and range of catching activities continued to diversify and distant-water fleets were soon exploiting virtually every known stock of fish across the open seas of the world. Over a period of twenty years the world's commercial fisheries increased their catch by over 300 per cent: from about twenty million tons in 1950 to about sixty-five million tons at the end of the 1960s.[1] The sheer speed of this expansion took little real account of the ability of global fish stocks to sustain such a level of exploitation and outpaced international attempts to secure effective conservation measures.

As competition on fishing grounds increased, so the concept of freedom to fish on the open seas was increasingly questioned, especially by states which were heavily dependent on fishing for their economic well-being. The trend over the fifties and sixties was to extend territorial waters and fishery limits. This trend progressively excluded British and other distant-water fleets from many of their traditional grounds, resulting in a series of fishing disputes between Britain and Iceland in particular, known as the Cod Wars.

During the 1950s Britain's distant-water fleet still ranged over great tracts of the northern seas. A twenty-one day Icelandic trip meant a round journey of some 2,000 miles from the Humber, whilst a trip to and from the extinct volcanic outcrop of Bear Island, 250 miles beyond the North Cape of Norway, involved a voyage of

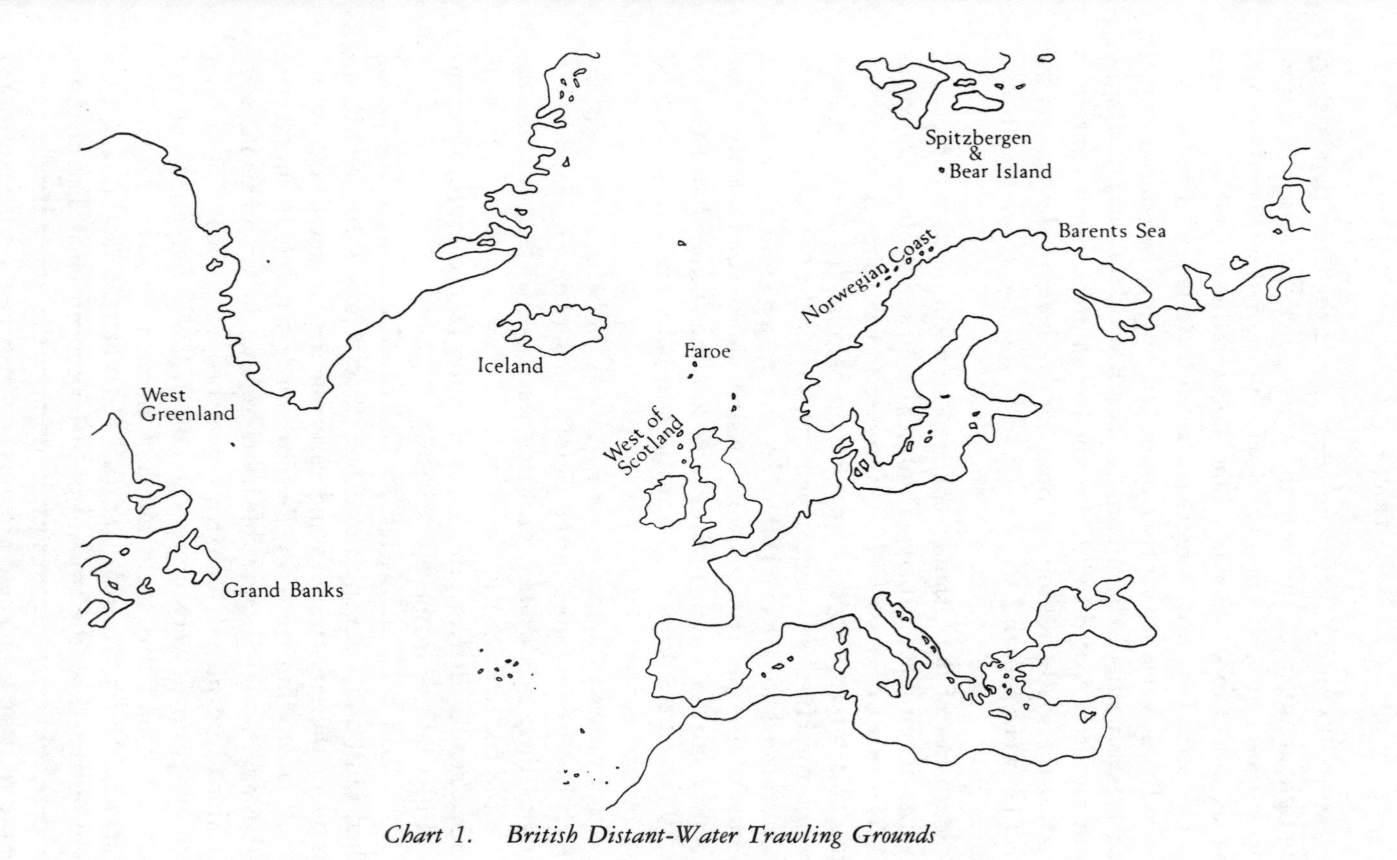

Chart 1. British Distant-Water Trawling Grounds

at least 2,700 miles. Though trawlers visited the Bear Island fishery throughout much of the year their main interest surrounded two concentrations of cod, the first of which occurred in May–June and the second October–January. The concentration of cod in the late spring attracted less attention because Icelandic fishing was generally lucrative at that time of the year anyway but the autumnal fishery was particularly important. The Bear Island fishery was the most northerly visited and during the late summer vessels sometimes reached the southern shores of Spitzbergen. Later in the year, as the weather worsened and sea ice spread south, the trawlers retreated before it.

South-east of Bear Island, separated by a 200-fathom depression, lay the Barents Sea fishery. Though the area was often known as the White Sea fishery no British trawlers in fact fished in the White Sea. This was the most distant of the regular fisheries and even though the Humber trawlers went by way of the Norwegian fiords they steamed for over 1,700 miles before lowering their nets. A round White Sea trip generally lasted about twenty-three days. About a quarter of all distant-water landings came from this area in 1954 and maximum landings were usually made in the second half of the year.[2]

Much closer to home was the seasonal Norwegian coast fishery which was concentrated in the vicinity of the Lofoten Islands. Most activity there was centred on a ten-week period from around the time the cod spawned in February, and only occasional trips were made after April. Nevertheless, the region ranked fourth in terms of Hull landings in the mid-1950s.

There had been an interest in Greenland ever since the period when Hull and Grimsby fitted out three factory ships for halibut lining in the late twenties and early thirties. A few distant-water vessels, including the *Cape Barfleur*, made trips there in the late 1930s and though grounds off Labrador and the Grand Banks were occasionally explored such trips stretched traditional side-fishing technology to its very limits. Because fish was only iced and not frozen, a round trip could not last much more than three weeks otherwise the catch could deteriorate too much. Moreover, long trips were restricted by the amount of fuel a conventional distant-water trawler could carry. Given such limitations the time a trawler could devote to fishing on these grounds was restricted by the additional length of the voyage. Thus only six or seven days might be possible in Greenland compared with ten or eleven in Iceland.

Despite these problems, the grounds could often be prolific. In January 1959 Bernard Stipetic, skipper of the new Hull trawler *Joseph Conrad*, owned by Newington Steam Trawling Company, set a new distant-water record when he made £18,965 from 4,384 kits (274 tons) of fish after a twenty-day trip to Greenland. But this had not been an easy trip for the *Joseph Conrad*, even by distant-water standards. During the voyage the trawler had been forced off two fishing grounds because of icing, and had steamed further north, finishing up on the Fyllas Bank. She was then delayed on the return trip by bad weather, battling through a hurricane round Cape Farewell, and arrived back in Hull almost out of fuel some two-and-a-half days late with her boat-deck rails flattened from the pounding she had received.[3]

In 1950 steam trawlers still dominated the distant-water fleet but the days of coal burning were numbered. Oil-fired boilers had been introduced just after the war when production problems in the mining industry made coal supplies unreliable. Several ex-naval patrol vessels were converted to oil firing when being refitted for trawling and the *Swanella*, ordered by J. Marr and Son in 1946, was the first purpose-built oil-fired trawler. The innovation, once accepted, spread rapidly. Though fuel oil was generally more expensive than coal a considerable amount of bunker space was saved and soon all new steam trawlers were being built with oil burners whilst many older vessels were converted. By 1955 only thirty per cent of Hull's fleet still burned coal and the port's last coal burner, the twenty-six year old *Othello*, was scrapped in 1963.[4]

Oil-burning steam trawlers continued to be built until almost the end of the 1950s for the trade was slow to accept diesel and diesel-electric trawlers. Though diesel engines were compact and efficient, allowing significant savings on space, they were initially more expensive to install and in the early days there was also a shortage of skilled diesel engineers. More importantly, perhaps, a diesel required many auxiliary motors as it had no steam to drive the trawling and steering gear nor the cod-liver oil plant. Though the first diesel distant-water steam trawlers were introduced in the late 1940s, Hull had only five such vessels as late as 1955. From then on, diesel gradually replaced steam.

The inshore fleets had used motor power as well as steam for many decades but, in Scotland at least, these motors had generally been installed in old sailing boats which had often been at sea since before the turn of the century. As the Government sponsored modernization programme got under way in the 1950s, many of these vessels, along with small steam trawlers and drifters, were replaced by new wooden vessels turned out by small yards under the new grant and loan schemes operated by the Herring Industry Board and White Fish Authority.[5]

Many vessels—whether inshore, middle- or distant-water—were able to benefit from wartime technological developments. In the post-war decade a range of electronic aids, including radar and sonar were becoming available to fishing boats. Soon they were able to plot their positions with more accuracy, locate shoals of fish in the water and even trawl around obstructions on the sea bed. Research and development in the chemical industries gave the fish trade access to a range of new synthetic fibres such as nylon, courlene and terylene which displaced natural fibres including hemp, cotton and manila in the manufacture of ropes and twines.[6] Nets made from synthetic fibres were far stronger and less easily torn than their cotton counterparts.

British distant-water landings reached their post-war peak in 1956 when the fleet brought no less than 8.5 million tons to market. Thereafter, fortunes followed a fluctuating but generally downward trend until 1964 when landings fell to 6.5 million tons—the lowest since 1935. But there was little fall off in the total availability of white fish with annual supplies generally hovering within three or four per cent of 1938 figures. Whilst the distant-water proportion of total demersal supplies fell from fifty per cent to forty per cent between 1956 and 1964 the shortfall was made up by an increased volume of near- and middle-water landings as well as an increase in imports which by 1964 were some 300 per cent higher than they had been in 1938.[7]

Though the near- and middle-water sectors were improving their performance this was not due to trawling. The increased landings could be attributed to modernization and the spread of seine netting. Back in 1938 seining had provided only two per cent of total landings from British vessels but the proportion had risen to seven per cent in 1951 and nineteen per cent by 1964. Whereas trawling had provided not less than ninety-two per cent of the total British white fish catch in 1938 by 1964 its contribution had fallen to seventy-eight per cent.[8]

The overall supply statistics also masked a declining consumer interest in fish products. Although total supplies of all types of fish remained close to their pre-war levels, the British population had continued to grow. Thus consumption per head had actually fallen by fifteen per cent as people became more affluent and a wider variety of foodstuffs became available. People were eating more meat, poultry, eggs, cheese, sugar and vegetables but less fish. Moreover, most fish eaten was still fried and few alternative ways of preparing fish were gaining ground. Aggressive marketing with new products and presentation was needed but there were serious obstacles to be overcome. Frozen fish seemed an obvious way forward but many people had been put off by the inferior products that had been forced on them during the war. Moreover, the British were always conservative in their choice of fish and continued to avoid 'new' or lesser-known species.

Table 8. Demersal Fish Landings by British Vessels in England and Wales from Selected Grounds. (cwt)

	1947	*1951*	*1956*	*1961*	*1966*	*1971*
English Channel	167,802	116,031	94,459	76,784	68,000	101,000
North Sea	1,865,823	1,313,596	1,225,239	1,481,407	1,851,000	2,269,000
Faroe	520,475	558,746	359,281	216,270	219,000	168,000
Iceland	1,245,531	2,618,721	2,937,119	2,978,902	2,685,000	3,373,000
Barents Sea	2,495,090	2,943,315	1,678,749	1,681,420	1,152,000	1,070,000
Bear Island/ Spitzbergen	1,805,459	1,871,421	2,911,322	1,178,570	354,000	60,000
Greenland (east & west)	1,781	34,948	65,060	140,417	375,000	45,000
Norwegian Coast	1,102,838	666,397	781,117	589,062	1,075,000	825,000

Figures from 1966 and 1971 rounded to nearest thousand.
Source: Sea Fisheries Statistical Tables.

Apart from disputes with Iceland, Norway and even Russia over fishery limits, the main reason for the declining performance of distant-water trawling lay in falling yields. The fall in distant-water landings prior to 1964 derived almost wholly from the Barents Sea and Bear Island fisheries. From a peak of 4.8 million cwt in 1951 these grounds supplied only 1.5 million cwt in 1966. Fish stocks there were being depleted and the catch per unit of trawling effort fell between 1955/6 and 1964 by about forty-five per cent in the Barents Sea area and by fifty-seven per cent at Bear Island.[9] Landings from the Norwegian grounds exhibited more stability but there was an increase of interest in Icelandic grounds—to the aggravation of the Icelanders. In the past, trawling companies had responded to such problems by increasing the efficiency of vessels and seeking out new grounds. Given this type of thinking, greater exploitation of north-west Atlantic waters seemed the obvious answer but it was already apparent that traditional trawler design had severe operational limitations when working so far from home.

One possible way forward lay with the introduction of freezing at sea and the development of factory trawling. Unlike the conventional fresh-fish trawler, a factory fishing vessel which caught, processed and froze its catch at sea would not be limited in length of voyage by the rate at which that catch deteriorated. It would be able to stay on distant grounds for a considerable length of time and spend proportionally far less of its working life voyaging to and from port. The concept of factory fishing had, of course, been pioneered in the late 1920s by Hellyer's of Hull and Bennet's of Grimsby with the *Arctic Queen, Arctic Prince* and *Northland* which froze line-caught halibut, but these vessels had discontinued their activities in the 1930s. After the war many experts were convinced that freezing at sea was an uneconomic proposition not merely because of the projected cost and scale of such operations but also because of consumer resistance to frozen fish. Thus traditional trawling firms showed little interest in factory fishing during the war. Yet the concept was still developed in Britain. The initiative, promoted initially by Sir Charles Dennistoun Burney, was taken up by Christian Salvesen & Co Ltd of Leith. Salvesen's were world leaders in factory ship whaling but whale stocks were diminishing and they sought to apply their formidable catching and processing expertise to fishing.

Factory whalers were large processing ships capable of undertaking long voyages between landings. They were also equipped with stern ramps for hauling up whales and Salvesen's believed that if trawlers were fitted out in a similar fashion they would be able to haul their gear and secure their catch while steaming head to wind. A conventional side trawler was obliged to stop dead in the water and turn broadside to the sea when recovering her trawl so that the force of wind and waves pushed her steadily downwind thus stretching out her trawl abeam where it would not foul rudder or propeller. Head to wind would make for a steadier operation and less manual effort as the net could be hauled up the ramp by winch alone. Moreover, the work of sorting and gutting the catch could be carried out under cover instead of out on an open deck exposed to the worst extremes of Arctic weather.

But the concept appeared to have one fundamental flaw: with a large catch there was always a danger that the cod end would split if brought over the side in one go. There was the additional danger that fish would be damaged by the sheer weight of their numbers. This was not usually a problem for side trawlers for they had no need to take the net aboard in one piece. Once they had secured the otter boards and the rest of the fore part of the net they could use a noose-like rig, sometimes known as a strop. This noose could be tightened to split about four or five tons of fish in the rear of the cod end off from the rest. This section could then be hauled aboard and emptied before being heaved back over the side and the noose slackened. The ship then moved slow astern for a moment whilst the fish collected in the cod end once more. The stropping operation could then be repeated several times until the entire catch was on board. A stern trawler with nets trailing well astern as it moved forward into the wind would be unable to carry out such an operation and would be obliged to haul her nets and cod ends aboard in one piece.[10]

Early stern trawling experiments were carried out with the steam yacht *Oriana*, whose stern was specially modified, then the company converted the former Algeria class minesweeper, HMS *Felicity*, into a stern trawler. Renamed the *Fairfree*, she was fitted with an experimental freezer embodying the principles of both plate and blast freezing. Initially, a new type of trawl incorporating a radical design of otter board known as a parotter was used but it was soon found that the traditional otter trawl was the most efficient. A series of voyages during 1947 and 1948 gave Salvesen's considerable experience of stern trawling techniques and once teething troubles were overcome they were convinced that the operation would work on a much larger scale.[11] Consequently in 1953 they placed an order for a factory trawler with J. Lewis, a small Aberdeen shipyard.[12]

The new vessel, named *Fairtry*, was launched in 1954 and represented a radical departure from both conventional trawler size and design. At 2,600 tons and 280 feet in length she could carry a crew of eighty whilst the latest side trawlers then coming into service had a crew of twenty and an average length of 170 feet. Her high freeboard, clipper stem and white superstructure made her appear at first glance more like an ocean liner than a trawler. Crew quarters, all above deck, were indeed much better than fishermen usually expected. Four-man cabins and ample hot and cold showers were fitted throughout the ship and a cinema was installed in the crew's mess.

Fairtry was designed for stern trawling and the fishing net was drawn up her stern ramp to the upper fishing deck where it was emptied before the fish passed through the hatches into the fish pounds below. Down on the factory deck the fish was successively gutted, washed, headed and filleted before being quick frozen and stored in the low-temperature hold. *Fairtry* was capable of freezing fish at a rate of thirty tons a day and could carry 600 tons of frozen fillets. Her boilers produced up to fifty tons of cod-liver oil whilst the fish-meal plants could process twelve tons of offal a day.[13] A maximum of 100 tons of meal could be stored each voyage.

The vessel encountered a number of teething troubles during her early months at sea. Her most serious problem was ironically due to her formidable fish catching efficiency. As predicted, cod ends burst but gear also jammed whilst winches were strained by the sheer weight of fish being hauled aboard. A series of detail modifications—including a twin-tailed cod end—and changes in working practice were implemented which allowed the vessel to overcome these problems and Salvesen's were soon more than happy with their new ship's performance. *Fairtry* had a seventy-day cruising range and usually fished on the Grand Banks. She regularly returned from a round trip with a full fish-hold in half that time.

Salvesen's ordered two similar vessels, *Fairtry II* and *Fairtry III*, each one being built with detailed alterations based on their growing experience of factory freezers. The vessels were too large for conventional fishing ports and worked out of Immingham on the Humber. Though technically very successful they were extremely expensive to build and operate, and the traditional British fishing industry failed to capitalize on the world lead which Salvesen's had established. Part of the problem appears to have been the difficulty of readily overcoming the British prejudice against frozen fish. Nevertheless, the *Fairtrys* continued to operate until 1967 when they were taken out of service after a steep fall in the price of frozen fish.[14]

For the rest of the 1950s other British trawler owners continued to develop the conventional side fishing trawler whilst a number of other countries followed up the factory freezer concept. Whilst the British ate less fish in response to increased market choice and greater affluence their eastern European counterparts faced a far different situation. The centrally planned economies of many communist states were unable to deliver the wide range of goods, services and foodstuffs that western consumers took for granted. Fish seemed an obvious way of making up for the shortcomings of state-planned agriculture, and consumers, seasoned to queues and shortages, seemed likely to welcome any new addition to their meagre range of choice.

In the nineteenth and early twentieth centuries eastern Europe had relied on the British herring industry in particular for fish supplies. In the 1950s, faced by acute shortages of foreign currency and the wish to promote their own economies, Russia, Poland and Eastern Germany embarked on the rapid expansion of their fishing industries. Unfortunately, their programmes were based less on the ability of world fish stocks to sustain new levels of fishing and more on their philosophy of economic planning which involved the setting of certain production targets to be reached within a fixed number of years.

Immediately after the war Russia had concentrated on exploiting fishing grounds relatively close to home, such as those in the White Sea, Novaya Zemlya and Baltic regions. But in the fifties the eastern bloc began to look further afield particularly to the Pacific and Atlantic oceans. Because of the Cold War communist countries could not expect much help from the western nations which fringed the Atlantic so they embarked upon a programme of expansion based upon the creation of fully integrated fishing fleets, capable of independent operation on the high seas. Such an initiative required the creation of a new range of fishing vessels and Salvesen's efforts in this direction attracted considerable attention.

Late in 1953 the firm building the *Fairtry*, J. Lewis of Aberdeen, received inquiries from the Russians about the possible construction of twenty-four factory trawlers of similar design. Lewis's were a small firm and an order of such size would have caused them considerable problems. Nevertheless, they began talking to the Russians and eventually, after some urging from a Conservative Government keen to increase eastern bloc trade, they agreed to lend plans of the *Fairtry* to the Russians for preliminary study. No order was ever placed with the British company but the Russians soon ordered twenty-four factory trawlers from the West German Howaldtswerke yards in Kiel. In 1956 the first two of these vessels, *Pushkin* and *Sverdlovsk*, began fishing on the Grand Banks. In design the *Pushkin* looked very like the *Fairtry*, and some naval architects later described the ship as a 'slavish copy'.[15]

With the aid of state subsidies the Russians rapidly built up a massive distant-water fishing capability by adding to the range and diversity of their vessels. Other craft included the 507-ton *Okean* class of side trawler, and later the 3,650-ton *Nekrasov* class of factory freezer was introduced. When at sea the fleet was largely self-sufficient, being serviced by oil tankers and water carriers whilst workshop vessels and ocean-going tugs were on hand to provide assistance. Transporters carried bags of fish meal and traditional barrels of salted fish as well as frozen fish blocks. Large mother-ships were to become a feature of the big fleets, being fitted out with hospital wards, operating theatres, cinemas and libraries as well as processing plants. Unlike the British fishing industry women were also employed on trawlers.

By 1960 the Russians had a large fleet working across the north-west Atlantic area. They not only fished on grounds long frequented by Spanish, Portuguese and French fishermen but also exploited recently discovered stocks of ocean perch off the coasts of Newfoundland and Labrador. Moreover, they continued to diversify their effort and were soon seeking species in the central Atlantic area as well as building on similar developments in the Pacific. As early as 1960 they were landing over three million tons of fish and had doubled this figure by 1968.[16]

Their efforts were emulated by the East Germans and Poles. The East Germans had little in the way of a traditional fishing industry but after the war developed an expertise in the construction of fishing vessels. In the early 1950s their fishing efforts were largely confined to the Baltic but then they increased their interest in the North Sea and by the end of the decade were working off the North Cape of Norway, Spitzbergen and moving into the north-west Atlantic. In the early 1960s they became interested in grounds off South Africa and South America and by 1965 were taking 300,000 tons of fish compared with 15,000 tons in 1951.

Poland, too, had traditionally relied on fish imports but during the 1950s considerably expanded the state-owned fleet. By the end of 1959 Poland possessed sixty-four steam trawlers and motor trawlers in addition to fifty drifter trawlers. State-owned enterprises had been gradually taking a larger stake in the industry and by 1959 owned 76.6 per cent of the fleet and accounted for more than three-quarters of all landings. Poland introduced factory trawlers with stern ramps in 1960 and by the end of the decade was well-established as a fishing nation taking over 400,000 tons of fish annually compared with just 46,000 tons in 1946.[17]

Whilst eastern Europe was building up its fishing industries a number of other nations, most notably Japan and Peru, were similarly expanding their catching capacity. During the 1950s Britain lost ground in the league of leading fishing nations, though Hull and Grimsby remained amongst the largest fishing ports in the world. With the exception of the three *Fairtry* vessels Britain's deep-sea effort was still based on the side fishing trawlers. Over the decade there had been a trend towards larger, if fewer, trawlers and though potential overall catching capacity was not much altered the distant-water fleet fell in size.

Until 1956 the British distant-water sector had generally paid its way.[18] But during the later fifties the problems associated with the Icelandic fishery dispute, poor catches and low dock-side prices made the trade more and more unprofitable. A Government Inquiry, the Fleck Committee, was set up to investigate the whole industry and following its recommendations the Government extended the near- and middle-water system of operational subsidies, grants and

Plate 8. The First Hull Stern Freezer Lord Nelson *on the River Humber*

loans to distant-water vessels. Operating subsidies were to be reduced progressively on the assumption that the sector would reach viability within ten years.

This financial package assisted owners in the development of a new marque of fishing vessel which, though somewhat smaller in scale, was broadly based on the *Fairtry* designs and enabled more intensive exploitation of the very distant grounds. In 1955 the trawlers *Keelby* and *Northern Wave* undertook trials of a 'robustly designed' vertical-plate freezer developed by the Torry Research station and jointly financed by the Government, the Distant Water Owners' Development Committee and the White Fish Authority. *Northern Wave* made eight voyages in all between December 1955 and July 1956 and successfully proved that the design could be developed for vessels somewhat smaller than the *Fairtry*. Later trials undertaken by Marrs of Hull and Fleetwood aboard the *Marbella* and *Junella* also showed that the crew could successfully handle the equipment. These vessels were all conventional side trawlers but in 1959 the first British stern trawler of more conventional size, the

104-foot *Universal Star* of Aberdeen, was built for middle-water trawling and in 1961 Associated Fisheries of Hull introduced the first distant-water stern trawler.[19] She was the *Lord Nelson* and was built at Bremerhaven in West Germany. At 238 feet she was about thirty feet longer than any conventional trawler and was fitted out with air-conditioned cabins as well as bathrooms and showers.

The *Lord Nelson* was part freezer, part wet-fish trawler, as only part of her catch was frozen at sea. Marrs completed their experimental freezing programme and ordered Britain's first all-freezer stern trawler, the *Junella*. This 1,435 ton (gross) vessel measured 238 feet from stem to stern and was launched in July 1962 from the Hall Russell yard in Aberdeen. *Junella* was powered by English Electric diesel-electric motors and her new vertical-plate freezers were designed to deal with twenty-five tonnes of fish a day.[20]

In the following year five of Hull's leading trawler companies placed orders for stern trawlers and the first of these, *Northella*, began fishing in September 1964. Grimsby's first stern freezer, the *Ross Valiant*, entered service the same year. The new marque soon proved its worth and no new distant-water side trawlers were ordered after 1964. By mid-1966 Britain had twenty-two stern freezers in her distant-water fleet and by 1969 the number had risen to thirty-three.[21]

Freezers had a different programme of operations from the side winders they were gradually replacing and the workload imposed on the crew was less arduous. They carried a crew of twenty-five or more, compared with twenty on the side-winders, including three deck officers. Though they spent up to seven weeks at sea they spent longer in port between trips. The fishing day was also marginally shorter, for deck crews were expected to work twelve-hour watches followed by six hours off which gave each man the equivalent of eight hours rest in every twenty-four. The crew were also able to spend less time in the open as the work of gutting and stowing took place under cover, though it was still necessary to be on deck for hauling, shooting and mending the nets. The less arduous working conditions on freezers were reflected in their health and safety record which was consistently better than on side trawlers.

The arrival of the freezer trawlers also changed the working arrangements of the men who unloaded the vessels. Traditionally, wood and metals kits were filled with loose fish and crushed ice by the bobbers of Hull and lumpers of Grimsby when discharging a side-winder but the new vessels landed frozen blocks of fish which passed by conveyor belt on to lorries waiting by the quayside to take the catch to the cold stores.

Many established firms in Hull and Grimsby invested in stern trawlers. Marrs, Hellyer's and Boston Deep Sea Fisheries all acquired vessels from the mid-1960s. Thomas Hamling's first such vessel, the *St Finbarr*, entered service in 1964. The *St Finbarr* lasted little more than two years before being lost after a blaze broke out in the crew's quarters on Christmas Day 1966. Twelve members of her twenty-five-man crew perished and although the stricken vessel was taken in tow by the *Orsino*, she sunk two days later off Newfoundland after the tow line parted. Fire seems to have been a regular hazard for the stern freezers. Indeed, several Hull and Grimsby vessels were lost or damaged by fire in the 1960s and 1970s.[22] The most mysterious freezer loss was that of the Hull vessel *Gaul* which disappeared in February 1974. Rumours have continued to persist that she was a spyship. Whilst it seems inconceivable that British Intelligence services did not make some use of British trawlers going about their legitimate business in sensitive waters close to the USSR, it seems most unlikely that such activity was linked to her loss. The official verdict that the *Gaul* was overwhelmed by a freak wave in the Arctic Ocean seems the most likely explanation.

The stern freezers were formidable fish-catching machines and the marque continued to be improved and developed into the 1970s. In 1973 Boyd Line had the *Arctic Buccaneer* constructed at Gdynia, Poland. Her first years were marred by technical problems and her maiden voyage was not made until 1975. She was designed as a vessel which could operate anywhere—in the Tropics as well as the Arctic—and possessed a fish hold capacity of 40,500 cwt. Another new vessel was the *Princess Anne*. Built for Boston Fisheries in 1974 in Wallsend, she was the only British-built trawler constructed to Lloyds Ice Class 1. In 1975 she took the British record for the most fish caught in one year.

Though the stern freezers were technologically the most advanced fish-catching machines in regular operation and considerably outclassed the old side-winders in terms of both range and catching efficiency, in reality they represented a very traditional trade response to changing circumstances. As we have seen throughout this book, the trade's usual way of responding to problems of maintaining or improving fish landings was to increase the catching efficiency of the vessels it deployed and seek out new grounds ever further afield. This process, begun by the nineteenth-century smackowners, was carried through with the development of the stern freezers to the end of the third-quarter of the twentieth century. But though the stern freezers gave the trade the means of exploiting grounds beyond the limits of the side-winder, the world was rapidly running out of such grounds, especially those with stocks of the traditional varieties of caught fish. Moreover, the extension of limits to 200 miles in 1977 effectively deprived the stern freezers of many of the open sea grounds they had started to exploit.

16

Cod Wars and Common Fishery Policies: The Beginning of the End

At the end of the Second World War both the British Government and the distant-water trawling industry were convinced their interpretation of territorial waters and fishery limits was enshrined in international law. Indeed, it was generally accepted that territorial limits stretched for no more than three miles from the low-water mark except in the case of bays less than ten miles in width where the three-mile limit was measured from a straight line drawn between two headlands. A state's fishery limits were taken to cover the same area. These principles had been adopted throughout the British Empire and agreed to by many other governments of coastal states by 1914, often through the medium of conventions with the United Kingdom. The Anglo-Danish Convention of 1901 which addressed the issue of limits around the Faroe Islands, Greenland and Iceland was one such treaty. A 1930 League of Nations conference found three miles acceptable to more countries than any other limit. Even so, the three-mile limit had never gained universal acceptance and a range of widths were claimed. Russia, for example, had claimed a twelve-mile territorial sea since 1909 and after a dispute with Britain an agreement was reached in 1930 by which British trawlers were allowed to fish as close as three miles from the coast or nine miles inside the Soviet definition of territorial waters. Some Mediterranean states had a six-mile tradition whilst Sweden and Norway had never really abandoned the old Scandinavian concept of a four-mile limit. By the end of the Second World War it was becoming clear that the seas contained valued and exploitable resources other than fish which complicated and widened the issue.

The war had highlighted the strategic importance of adequate oil supplies and the United States Government was particularly concerned about the future exploitation of offshore oil reserves and other mineral resources as well as fish stocks. In 1945 this concern manifested itself in two declarations which became known as the Truman Proclamations. One of these claimed jurisdiction for the USA over natural resources in and on the continental shelf whilst the other claimed the right to establish fishery conservation zones across areas of the high seas contiguous to her coasts. A number of other states, including Mexico, Chile and Costa Rica, followed the US example during the next few years whilst Peru claimed control and protection over a 200-mile zone of the continental shelf from August 1947.[1] Such moves undermined the general validity of the three-mile limit.

But across the Atlantic, Norway had reasserted the Scandinavian concept of the four-mile limit in the 1930s. Her claims for an extended limit had caused little international concern until 1906 when British trawlers started working off her northern coasts. A number of incidents occurred over the following years and on 12 July 1935 a Royal Decree was issued which laid down the limits of the northern coast fishery zone from Varanger Fjord to Vestfjord.[2] This not only confirmed the four-mile limit but also differed markedly from the method by which the United Kingdom believed limits were determined under international law. Instead of taking measurements from the low-water mark, the 1935 decree identified some forty-eight fixed points on either the mainland or islands as well as rocks or sea-stacks, some of which were a considerable distance from the mainland. These points were connected by straight lines, known as baselines, between eighteen and forty four miles in length, and the sea enclosed by them, as well as the adjoining four-mile belt of water, was said to constitute Norwegian limits.[3]

The British had originally objected to the decree but negotiations on the issue had made little progress by the outbreak of war. Afterwards, British trawlers renewed their interest in the disputed area and the number of incidents increased. Many trawlers were arrested in 1948/9 and the United Kingdom Government instituted proceedings against Norway before the International Court of Justice in The Hague. The British side, led by the Attorney General Sir Frank Soskise, did not take issue with the four-mile limit but contended that the Norwegian use of baselines was contrary to international law. The Norwegians for their part pointed out that protection of local fishing interests was especially important in this area north of the Arctic Circle, being the traditional and main means of existence for its inhabitants. They also argued that they had never accepted the British method of determining limits, which was in any case extremely difficult to apply due to the rugged and indented nature of the north coast. Their system, they claimed, was well established by constant and long practice dating back to at least 1812 and had encountered little opposition until relatively recently. Even the United Kingdom had not made a formal protest before 1933.

In December 1951 the International Court found in favour of the Norwegians by ten votes to two. It accepted their arguments on geographical and economic difficulties as well as long-established practice. In effect it rejected the open-bay rule, one of the two main principles of the three-mile system, by ruling that Norway's system of drawing straight baselines was not contrary to international law, even though it meant the closure of all bays. Though Britain accepted the ruling it was taken aback and with good reason, for though a regional dispute, it opened for question the whole concept of fishing limits and territorial waters under international law. Britain and many of her neighbours thought such issues had long been settled by bilateral treaties or established custom and practice.[4]

Iceland viewed the issue with far more than academic interest. Though it was sometimes argued that Iceland would be rendered uninhabitable if the fish stocks off her shores were destroyed, the country was in fact richly endowed in certain other natural resources, including geothermic heat and almost unlimited capacity for generating electricity through hydro-electric power.[5] Nevertheless, fishing was vital to her economy and was the principal means of earning foreign currency with which to develop her other resources. During the 1950s fish and fish products made up around ninety-five per cent of all Icelandic exports.[6] After the foundation of the Republic of Iceland in 1944 and the severing of the last colonial links with Denmark, this small nation of 230,000 people was keen to protect what she saw as her crucial interest in the fish-rich waters around her coasts.

There had been keen discussion in Iceland during the inter-war years about the possibility of banning all fishing in the breeding and nursery grounds of Faxa Bay but these came to nought. In 1938 Iceland had become an independent member of the International Council for the Exploration of the Sea and in 1946 the Council finally passed a resolution recommending the closure of Faxa Bay to trawling and seine netting as an international experiment for a period of ten years. With a view to negotiating a treaty on this issue Iceland invited all interested states to a Faxa Bay conservation conference. But although a few countries agreed to participate the British Government, whose trawlermen took the greatest share of the catch there, declined to participate and Iceland reluctantly abandoned its conference proposals.[7]

During the war, when British trawlers had been requisitioned for military service, Iceland had been encouraged to expand her own fishing industry but afterwards British and other European trawlers returned to the grounds in ever-growing numbers. Despite the disruption of distant-water fishing during the war there was a long-term tendency towards increased exploitation. Over the period 1937–1954 the total catch of fish from Icelandic waters more than doubled. Iceland's own share rose from 31 per cent to 48 per cent and though the British and West German catch took a declining percentage of this growing overall catch they both increased their total catches.[8] In the immediate post-war period British trawlers visiting Iceland were rewarded by bumper catches after nearly six years of under-exploitation but evidence of over-fishing soon appeared.[9] Though cod catches held up at first between 1949 and 1952, total landings of haddock and plaice from Icelandic waters were to decline by 32.5 per cent and 47.3 per cent respectively.[10] To the Icelanders these figures seemed confirmation of their worst fears and justified the unilateral moves they began in the later 1940s.

Faced with an apparent lack of foreign will—particularly on the part of the British—to understand the strength of her interest in conservation measures, a determination grew within Iceland to take control of issues concerning her continental shelf. On 5 April 1948 Iceland passed the Law Concerning the Scientific Conservation of the Continental Shelf Fisheries. This law, which in parts closely followed the Truman wording, enabled the Minister of Fisheries to determine conservation zones within the limits of the continental shelf wherein all fisheries would be subject to Icelandic rules and control.[11] The Icelandic Government tried hard not to contravene international law and the Act stated that it 'could only be enforced so far as was compatible with agreements to which Iceland properly regarded herself as bounded'. These, of course, included the 1901 Anglo-Danish Convention.

In 1949 Iceland informed Britain of her intention to terminate the 1901 Anglo-Danish agreement on her limits and, as two year's notice of abrogation was required, this duly expired in 1951. Meanwhile, in 1950, regulations setting out a four-mile limit on the north coast from a baseline drawn through twelve points were issued on the basis of the 1948 Conservation Law. These regulations prohibited trawling and seining within these new limits by Icelandic and other nationals except where they conflicted with international agreements. As the old Anglo-Danish Convention was not due to expire until 1951 this meant that British trawlers could continue to work for a time in areas prohibited to Icelandic and other vessels.[12] By the time that Convention expired in 1951 the Anglo-Norwegian fishery limit dispute was pending at the International Court. After the decision went in favour of Norway, Iceland felt justified in setting four-mile limits from straight baselines across bays. Regulations to that effect were subsequently issued and came into operation from 15 May 1952.

The new limits, extended from three to four miles, were determined from baselines drawn from fifty-one points on islands and rocks off the coast as well as promontories. Within them all trawling and seining was prohibited. Britain, Belgium, France and Holland all protested but the British reaction was the most vehement and sustained. Apart from protesting about the baselines, especially as they closed off Faxa Bay, Britain argued that Iceland did not have the right to unilaterally extend limits, and apart from depriving trawlers of the right to fish on what she saw as the high seas no account was being taken of historic usage. Iceland for her part contended that neither the three-mile limit nor the ten-mile rule in bays had any authority in international law. She further claimed the right to determine the extent of her jurisdiction over coastal fisheries in view of economic and biological considerations.

Sections of the British trawling industry were incensed by the Icelandic move and pressure was soon being exerted for a trade ban on Icelandic landings, which many thought would soon bring Iceland round as Britain accounted for twenty-five per cent of her exports. In October 1952 representatives of the British Trawler Federation met their Icelandic counterparts but many trawler owners, officers and crews took an unyielding stance on the issue. The Grimsby Trawler Officers Guild was especially militant and threatened strike action should Icelandic landings be allowed. Indeed, a total ban was finally imposed by Hull and Grimsby merchants in November 1952 after a three-day strike on the issue.[13]

The dispute dragged on for over four years but banning Icelandic landings did not have the intended effect. Ironically, it may well have strengthened the Icelandic economy by forcing them to diversify fish export products and markets. Initially, Iceland found a ready market in Italy but in the longer term she invested in more freezing plants and found growing markets for frozen fillets in both the USA and Soviet Union in particular. New markets were also opened up in Brazil, Cuba and West Africa.[14]

There were several attempts to get the issue referred to the International Court at The Hague but the two sides could never agree on the terms of referral. The matter was eventually settled through the auspices of the Organisation of European Economic Co-operation and Development on 14 November 1956 and two weeks later Icelandic landings resumed when the trawler, *Ingolfur Arnason* brought 34,000 stone into Grimsby.[15] Though Iceland agreed not to extend her limits, pending further discussions by the UN General Assembly of the report of the Law of the Sea by the International Law Commission, and though the UK reserved its position regarding the legal validity of Icelandic methods of determining limits,[16] Britain and other states in effect recognized her new four-mile limit.

In Iceland the political and economic arguments in favour of a further extension of control over neighbouring fish stocks were strengthened from the mid-1950s by more indications of over-fishing. Between 1954 and 1957 the total annual demersal catch in Icelandic waters fell from 881,147 tonnes to 743,316 tonnes even though fishing effort, in terms of vessels and nets deployed, was increasing. The average yield of demersal species per 100 hours fishing by Belgium and British trawlers fell by between thirteen per cent and twenty-one per cent over the same period.[17] Conservation issues clearly needed addressing and Iceland, with her economy so dependent on fishing, was determined that foreign fishermen, who still took more than half of the catch, should face further restrictions on access.

Back in 1949 the Icelandic Government had successfully proposed to the United Nations General Assembly, in the face of strong west European opposition, that the newly appointed International Law Commission should include the issue of territorial waters in its priority study list. The International Law Commission subsequently spent several years considering the issues of high seas and territorial waters and subsequently reported back to the UN General Assembly in 1956. The Icelandic Government hoped to press home another extension of her fishery limits on the basis of an international agreement after the UN had considered the International Law Commission's report. However, the General Assembly felt the matter was too complex to be dealt with directly and only Iceland was to vote against a decision to refer the matter to a conference on the Law of the Sea.

This first UN Conference on the Law of the Sea met at Geneva between February and April 1958. By this time there had been a great many states making a diverse number of claims to waters off their coasts well beyond the old three-mile limit. At first Britain fought a rearguard action in support of three miles whilst Iceland, in contrast, favoured a twelve-mile territorial and fishery limit. When it was clear her position was untenable, Britain went on to support a proposal for a six-mile territorial and fisheries limit but ultimately supported a US proposal of a six-mile territorial limit surrounded by a further six-mile exclusive fisheries zone in which nations with historic rights could continue to work for five more years.[18] This last proposal won most favour but failed to obtain the requisite two-thirds majority. The joint twelve-mile territorial water and fishery limit proposal favoured by Iceland received much less support. The conference went on to recommend that the UN should consider calling a further world conference on the issue.

Because fishing was so vital to its economy the Icelandic

Government came round to the view that the country had a special case and should have the right to the widest of fishery and territorial limits. It was deeply dissatisfied by the failure of this first Law of the Sea Conference to agree on new limits and was determined not to wait until a second was convened before implementing any further extension. Therefore, on 30 June 1958 the Government used the 1948 Conservation Law to issue a decree extending fishery limits to twelve miles from 1 September.

Britain and many of her neighbours, including Belgium, Holland, West Germany and Spain protested strongly and when representatives of the Western European fishing industries met at The Hague on 14 July 1958 they not only passed a resolution condemning the Icelandic action as illegal but also declared their intention of continuing to fish up to the four-mile limit. Informal talks between both sides were held under the auspices of NATO and although compromise proposals were apparently put to Iceland they came to nothing. In the event, all the protesting states respected the new limits from 1 September with the exception of the United Kingdom and the ensuing dispute, otherwise known as the First Cod War, was to last until 1961.

In Britain both trade and Government were determined that fishing would continue in the disputed waters and plans were laid to provide Royal Navy protection. Throughout August trawler skippers between trips were secretly briefed about strategy by owners and navy alike.[19] Initially, the forty or so trawlers fishing in the disputed waters operated in one of three packs, each guarded by a naval vessel on protection duty. Within a few hours of the decree coming into force Icelandic coastguard vessels made at least two attempts to intercept Grimsby trawlers. A couple of days later, under the cover of fog, the Icelandic coastguard vessels *Thor* and *Maria Julia* closed on the Grimsby trawler *Northern Foam* and sent across an unarmed boarding party. Skipper J. Crockell refused their order to leave his vessels whilst the radio operator locked himself in the cabin and, despite Icelandic attempts to jam his broadcasts, succeeded in summoning the frigate HMS *Eastbourne*. As the frigate closed the Icelanders tried to take *Northern Foam* away at full speed but her crew immobilized the engines. *Eastbourne* sent over her own boarding party and retook the trawler. Although there was no violence throughout the incident, the *Thor* refused the British request to take the boarding party back and the frigate was left with nine unwelcome guests.[20] They were eventually set loose one night in the *Eastbourne*'s motor cutter near the Icelandic coast: no more Icelanders were taken during the conflict.[21]

This incident more or less set the scene; throughout the conflict the Icelanders harried and attempted to arrest British trawlers fishing within the designated protection boxes whilst the Royal Navy constantly tried to frustrate them. Quite a number of trawlermen had seen action during the war in the Patrol Service and knew more than a little about defending themselves. Iceland, in return banned Royal Navy vessels from ferrying sick or injured fishermen ashore. This had serious implications; if trawlers landed their own sick in Icelandic ports they risked arrest if they had been seen fishing within the twelve-mile limit. As the dispute wore on it grew in bitterness, even occasionally becoming violent. There were numerous 'collisions' or 'rammings', and a whole string of related incidents. British trawlermen were accused of resorting to hosepipes and, on at least one occasion, iron bars and axes to repel boarders whilst the Icelanders were in their turn accused of a range of acts including the drawing of pistols and the use of sandbag coshes during an unsuccessful attempt to arrest the Grimsby trawler *Northern Queen*.[22] The trawler *Grimsby Town* seems to have been heavily involved in the action as late as the summer of 1960, claiming to have been subjected to a flare attack by aircraft and, a few months later, being chased and shot at by the Icelandic coastguard cutter *Odin*. On this occasion *Grimsby Town*'s skipper turned on his assailant and tried to ram her.[23]

Public opinion in Iceland was strongly behind the extension of limits and the offshore confrontation aroused strong passions. On one occasion the British Embassy in Reykjavik was stoned by an angry mob. Fire crackers and smoke bombs were also reportedly thrown whilst the British Ambassador, Sir Andrew Gilchrist, played classical music at volume to keep up staff morale.[24] In Britain sections of the trawling trade again tried to secure a prohibition on landings by Icelandic trawlers and even attempted industrial action, but this time Icelandic landings were not banned. The door of negotiations was left slightly ajar.

During the dispute it has been estimated that the Royal Navy deployed at one time or another at least thirty-seven different warships, including about seventeen destroyers and nineteen frigates able to make speeds of between twenty-two and thirty knots. Iceland, in contrast, fielded six small vessels, the largest of which was the coastguard cutter *Thor* (700 tons and twenty-eight crew), which could manage between eleven and seventeen knots flat out.[25] However, the British trawlers and naval escorts were operating far from home and in any case military might was not to be the determining factor. The dispute may have been between Britain and Iceland but like all the cod wars it acquired an international dimension. Iceland was a member of NATO and occupied a strategic position in the Cold War conflict. Cyril Osborne, Conservative MP for Louth, was not alone when he described Iceland as the Gibraltar of the North[26] and there was always the worry amongst NATO allies that the dispute might drive Iceland into the arms of the eastern bloc. Crucial to NATO's northern strategy was the US airbase at Keflavik—far from popular with Icelanders—and the fear was that this might be lost. Such factors bore heavily on the British Government already faced by mounting costs and lost fishing opportunities as the dispute dragged on and on.

In late December 1958 economic problems brought down the leftist Icelandic Government of Hermann Jonasson. A minority government led by Emil Jonsson took over, but by the end of November 1959 the country was being governed by a right inclined coalition of the Independence Party and Social Democrats led by Olagur Thurs. These two parties had been more moderate in their original approach to the dispute but, though perhaps more inclined to compromise, they were still against negotiation under what they saw as military duress.[27]

In the following spring a possible way forward appeared with the calling of the second UN Conference on the Law of the Sea. Before the Conference opened the British distant-water trade suspended fishing operations within the disputed waters as a gesture of goodwill.[28] But this Conference failed to agree on the crucial issues of territorial seas and fishery limits, though it soon became obvious to the British delegation that support for the old three-mile limit had all but evaporated. Britain eventually supported the US–Canadian compromise favouring the six-plus-six proposal which differed from that put forward at the last conference in that it would have given foreign nations with historic rights ten rather than five years fishing in the outer six miles. This proposal failed to gain the necessary two-thirds majority for acceptance by just one vote.

After the Conference had broken up the British Government announced that Royal Navy vessels would stay out of the disputed waters for the present time and the trawling trade told trawlers to work outside the disputed waters for a further three months; this was later extended into the autumn. Both countries exchanged views through the auspices of NATO and the Icelandic Government also waived all charges against British trawlers for earlier infringements of its twelve-mile limit. As relations improved the British Prime Minister, Harold MacMillan, met his Icelandic counterpart, Olafur Thors, at the Keflavik air base and their discussions set the scene for greater understanding.

The first round of negotiations between teams of officials began in Reykjavik in October 1960. Over the following few months, dialogue continued at various levels and their respective Foreign Ministers discussed the issue at a NATO meeting in Paris during December. Though this process was eventually to bear fruit it was not without its problems. The British felt that Iceland was prolonging the issue unnecessarily whilst, for their part, the Icelanders were aggravated by a British threat to return Royal Navy vessels to disputed waters.[29]

The dispute was finally resolved in March 1961 and Iceland secured British recognition of the twelve-mile limit with the proviso that British vessels would be able to fish to within six miles of the coast in certain areas at specified times of the year for a further three years. British pressure also secured an agreement that either party could refer any dispute over further extensions of territorial waters or fishery limits to the International Court of Justice.

Both Norway and the Faroe Islands were also to extend their limits to twelve miles, though they agreed to transitional concessions for British vessels. The British Government also faced pressure from inshore fishermen, particularly in Scotland, for similar limits, but it was keen that any home extension should be part of an international movement rather than a unilateral action. Britain eventually called a European Fisheries Conference, which met in London for three sessions between December 1963 and March 1964. The so-called London Convention emerged, enabling the signatories to regulate fisheries within a twelve-mile zone from baselines drawn from headlands. The inner six miles were to be reserved exclusively for native fishermen, subject to a short phase-out period for foreigners who had traditionally worked there. The outer six were also reserved for native fishermen but foreigners could continue to work the grounds and stocks they had traditionally fished.[30]

The main British beneficiaries were the inshore fishermen but the middle- and distant-water trade found little relief as Norway and Iceland did not accept the London Convention and Denmark would sign it only in respect of her own coastal waters and not on behalf of the Faroe Islands or Greenland.

The Convention was signed around the time that the agreement giving British vessels certain transitional concessions within the Faroese twelve-mile limit was coming to an end. The failure to secure Faroese acceptance of the Convention placed a number of traditional grounds out of bounds to British trawlers and this hit Aberdeen, Granton, North Shields and Grimsby's middle-water fleets particularly hard. The British trade retaliated by forming the British Fishing Industry Committee on Imports from the Faroes and this group imposed a annual quota of £850,000—equal to the previous ten year average of annual Faroese landings—on fish imports from the islands.[31] Though this eventually led to discussions with the Faroese fish trade, their government refused to back down on the issue of limits.

During the 1960s Britain became increasingly interested in joining the Common Market. At the time membership seemed attractive to the distant-water trawling trade for it offered the prospect of tariff-free access to the markets of the original six member countries. Though outside the Common Market, Britain was a member of the European Free Trade Association but to sections of the British fish trade this arrangement seemed to offer the worst of both worlds. Under the terms of the EFTA Convention Britain was obliged to lower import duties on a range of fish products including quick-frozen fillets. Denmark and Norway were fellow members of EFTA and, having large fishing industries, could only benefit from this arrangement. Thus the British fish market was being opened up to the very states which were pursuing policies aimed at closing British access to waters around their coasts or territories. At the same time it was becoming increasingly difficult to export to the Common Market as member countries raised their tariffs on outside fish products. This also raised Scandinavian interest in the British market.

Throughout the 1960s the Common Market lacked a broad fisheries policy. Under the Treaty of Rome fish had been defined as an agricultural product and the original fisheries policy covered little more than the reduction of tariffs between member states. It was not until 1966 that the Commission produced its first report on a wider policy and it took a further two years for substantial proposals to be presented to the Council of Ministers. Even then there was still little real agreement on the nature of a Common Fisheries Policy. However, the Community impetus for agreement quickly gained momentum when Britain, Denmark, Ireland and Norway seemed likely to be successful in their application for membership. All four states had important fishing industries and fishing policies. The prevalent view amongst the six existing members was that the Community should also be armed with a fisheries policy before negotiations opened and thus the broad principles of a Common Fisheries Policy (CFP) were hurriedly drawn up and agreed to on 30 June 1970, just one day before the membership negotiations opened. As a result, the original six members incorporated into Community law the principle of equality of access to the waters of all Community countries for all Community vessels.[32]

The manner in which the CFP was drawn up and presented, as well as the principle of equality of access, upset all the applicants. As recently as 1964 nine out of ten states involved—the exception, of course, being Norway—had signed the London Convention and accepted a definition of national limits which the CFP seemed to throw out in favour of allowing every Community member's fishing vessels the right to work up to the beaches of the others. The combined fishing industries of the four applicants greatly outweighed those of the six and possessed far-richer fishing grounds. It appeared to the applicants that the six were deliberately overriding the legitimate interests of the coastal states in order to guarantee their relatively small fishing industries access to previously exclusive fishing grounds. All applicants pressed strongly for revision but it was soon clear this would be difficult. In Norway this issue was undoubtedly the key reason why the voters rejected membership in September 1972. In Britain, however, it is clear that the Conservative Government was determined not to let this issue block entry. Though the inshore fishermen were angered by the CFP they were not, at the beginning of the 1970s, yet the dominant or most influential branch of the industry and the deep-water sector was less directly affected.

The agreement reached on the fisheries issue and incorporated into the Treaty of Accession allowed a ten year derogation from the equality of access principle. Until 31 December 1982 all member states could keep a six-mile fisheries limit around their coasts and a twelve-mile limit on agreed grounds providing fishermen with historic rights were allowed access. In Britain's case this twelve-mile agreement accounted for ninety-five per cent of her coastline and the Government claimed to have won a major concession, arguing that it would have the power of veto in 1982. However, their opponents argued that, being only a limited derogation, it would lapse in favour of the equality of access principle and the other states would have the power to veto any new arrangements.[33] The uncertainty this bred was a recipe for conflict within and between Community states for the rest of the decade.

Meanwhile, after the general election of 13 July 1971, the leftist coalition of Progressive Party, People's Alliance and Liberal Left Party came to power in Iceland's first major political change for over a decade. Though the previous coalition of Independence and Social Democratic Parties, which had made peace with the United Kingdom back in 1961, had recently passed a new law relating to continental shelf resources they had not been prepared for a further unilateral extension of limits. The new government, led by Olafur Johannesson, was far less reticent and on the very day it came to power it announced that the fishery agreements with Britain and Germany would be terminated and fishery limits would be extended to fifty nautical miles from baselines not later than 1 September 1972. Both Britain and Germany felt that Iceland had no right to take such action and three days later the British Ambassador delivered an *aide-mémoire* expressing his government's concern.[34] Over the following months both sides set out their positions, with Iceland once more making much of its unique economic reliance on fish stocks, and fears that these were increasingly in danger of being over-fished by an ever more efficient world distant-water fleet. She continued to take issue with the concept of freedom of the seas, arguing that the neighbouring continental shelf was an integral part of her territory and that fish stocks were as much a natural resource as was the oil and natural gas beneath the sea bed which so interested the British in the North Sea. The British Government, in return, argued that there was a clear distinction under international law between resources found in and under the sea and consequently denied Iceland's case on the conservation of fish stocks.

Efforts to solve the dispute over the ensuing months made little headway. Though the International Court of Justice made an interim ruling protecting British and West German interests, Iceland ignored this and extended her limits on 1 September 1972. While all other states kept their fishing vessels out of the disputed waters both British and West German trawlers continued to fish there. True to tradition the British trawlermen proved at their most defiant when their livelihood was threatened, and a number obscured their names and numbers in contravention of international law. Some even hoisted the Jolly Roger.[35] The Icelanders, for their part, adopted a new weapon, the trawl wire cutter, similar in purpose to the nineteenth-century 'devil'. Within a few days of Iceland's extension, her coastguard vessel *Aegir* used its wire cutter to sever the trawl warps of the *Peter Scott*[36] and the Second Cod War was underway.

With both sides determined to thwart the other, the wire cutting tactic inevitably brought Icelandic cutters and British trawlers into close contact and resulted in claims of deliberate ramming from both sides. The first serious incident occurred on 18 October 1972 when the *Aegir* and *Aldershot* collided during an apparent warp cutting attempt. The *Aegir* sustained ripped plating whilst the *Aldershot* was seriously damaged in the stern. Iceland afterwards claimed the *Aldershot* deliberately rammed the *Aegir*—presumably with its stern—whilst the trawler asserted that the Icelandic cutter had misjudged its warp cutting run. Though *Aldershot* was holed above the waterline the thrust of her propellor drove water in through her stove-in plates and the crew struggled to plug the gap with mattresses and matting whilst the trawler made its way to Faroe. After an unfriendly welcome from young demonstrators at Thorshavn they moved on to Skaale where they patched the stern and filled the gap in the plates with cement. Within 36 hours *Aldershot* was back at work off Iceland.[37]

The Icelanders showed equal persistency and soon became formidably efficient trawl cutters. Fishing operations were frequently disrupted and the trawler skippers demanded better protection. After the trade increased pressure on the Government, four ocean-going tugs, *Englishman, Irishman, Lloydsman* and *Statesman* were chartered to provide protection.[38] These tugs, though unarmed, proved vigorous in their defence of the trawlers and the resultant clashes increased bitterness on both sides. Iceland continued to harry the trawlers; by the end of 1973 the warps of no less than sixty-nine British and fifteen German trawlers had been cut.[39] On 18 May 1973 the trawler skippers moved out of the disputed waters telling their employers that they would no longer return without naval protection. Two days later, after strong pressure from the Humber ports, several British frigates were despatched to the disputed waters. The trawlers began fishing there once more and the tactic of fishing from protective boxes patrolled by frigates and tugs was used. When Icelandic coastguard cutters approached these areas, frigates or tugs would steam towards them and steer a parallel course for up to several hours at a time. At such close quarters it often turned into a dangerous game of cat and mouse and, as vessels vied for advantage, sudden shifts of course inevitably led to collisions. Though the Icelanders claimed the Royal Navy was deliberately inflicting damage, they may well have gained from such encounters for their vessels, if damaged, could soon reach home port whereas the frigates were obliged to make the long voyage home for repair, thus adding to the already huge cost of the conflict.

After the Royal Navy was drawn into the conflict, the Icelandic Government banned British military aircraft from landing in its country, claiming they had been reporting on the movements of Icelandic coastguard cutters as well as their normal NATO patrol duties. They also demanded that NATO should try and bring about the removal of British warships from the disputed waters. From 20 September 1973 they banned the landing of sick and injured personnel from the British fleet unless brought in by their own vessels, which once more left trawlers open to possible arrest, But the conflict continued and Iceland finally threatened to break off diplomatic relations. An open breach between two allies would have suited neither side, nor NATO. Edward Heath, the Conservative Prime Minister, invited his Icelandic counterpart, Olafur

Johannesson, to London for talks and agreed to withdraw the frigates and tugs from the disputed waters from 3 October.

A tough round of bargaining in London followed before the basis for a settlement was drawn up which was accepted by the Icelandic Althing on 13 November 1973. The main provisions allowed the British restricted access to the disputed waters. All British factory and freezer vessels were excluded and only 139 wet-fish trawlers were licensed to work between twelve and fifty miles offshore, of which no more than sixty-eight could exceed 180 feet in length. A number of conservation and small-boat fishery areas were closed to all British trawlers and the remaining waters were divided into six areas, one of which would always be closed to British trawlers on a rotating basis.[40]

This interim agreement was valid for two years and based on an estimated catch of 130,000 tons as opposed to the 170,000 tons permitted by the 1972 International Court ruling which had been ignored by Iceland. The potential catch was estimated to be worth up to £7.5 million a year but the restrictions placed on the trawling fleet represented a reduction in the number of vessels by about twenty-five per cent. Iceland, however, was unable to reach a similar agreement with West Germany until after the Fourth Cod War.

Throughout the period of the Second Cod War world opinion appeared to be moving in favour of allowing coastal states control over a wide area of resources in and around the seas which surrounded them. The creation of two sea bed committees in the later 1960s had made the United Nations the focus once more for developing legislation in this realm and in 1970 the General Assembly agreed to convene its third Conference on the Law of the Sea. This met for the first time in New York in December 1973 and progress as always was slow, but by the time its third meeting ended in May 1975 it was clear that more than 100 states already supported the right of coastal states to establish 200-mile Exclusive Economic Zones. Moreover, the final negotiating text, issued in May 1975, specified that in a state's Exclusive Economic Zone—which could extend up to 200 nautical miles from the territorial sea baselines—it should determine both the allowable catch of living resources and its capacity to harvest them. It was obvious that the Law of the Sea Conference would take some years to finish its work but Iceland took the view that these principles were here to stay and would form an indispensible part of the final outcome. Her Government argued, however, that cod stocks were so depleted through over-fishing that this further extension of the fishing limit could no longer be delayed.[41]

Iceland had based all previous extensions on its 1948 Fundamental Conservation Law and had already taken the precaution of amending this to cover an extension up to 200 miles from its baselines. In May 1975, shortly after the Law of the Sea Conference had issued the single negotiating text, the Icelandic Government extended its limits to 200 miles. The measure came into force on 15 October 1975 but since the British agreement over the fifty-mile limit did not expire until 14 November 1975 the regulations were not at first enforced against British trawlers.

This Third Cod War soon seemed to be heading the way of previous conflicts for all trawlers left the disputed waters with the exception of the British and West Germans. The latter, however, were not backed up by protection vessels in the disputed waters and their Government was soon seeking a settlement, whereas Britain's Labour Government resorted to the tactics deployed by its Conservative predecessor during the previous Cod War. First, ocean-going tugs were sent in and these were soon followed by Royal Navy frigates.

Within ten days of the Icelandic coastguard cutters moving in, three British trawlers, *Primella, Boston Marauder* and *St Giles* had each lost £21,000 worth of gear when their trawl warps were cut, whilst the *Real Madrid* managed to retrieve hers after a single warp was severed.[42] In the early stages of this conflict they were only protected by ocean-going tugs and, being strung out over a wide area, were easy targets for the Icelandic cutters. Disillusioned skippers found their operations so disrupted that they threatened to pull out on 25 November 1975 unless they received better protection, and the British Government sent the Royal Navy in once more. Harassment continued, but although ten trawlers had had their warps cut by the end of the year, once they were assigned to protective boxes they were, by and large, able to fish effectively again.[43]

Between early December 1975 and mid-February 1976, Icelandic and British vessels collided no less than 17 times as the Icelanders continued their warp cutting policy. Iceland regarded these as

British rammings and in early January threatened to break off diplomatic relations. Dr Joseph Luns, Secretary General of NATO, was seen as a potential mediator and invited to Iceland. He subsequently visited London but his talks proved inconclusive. Nevertheless, after Iceland again threatened to sever diplomatic links, Britain withdrew her warships on 19 January 1976 and invited the Icelandic Prime Minister, Geir Hallgrimsson, to London. The subsequent talks ended in failure and on 5 February 1976 the Royal Navy once more returned to the disputed waters. On 19 February, Iceland finally broke off diplomatic relations.

Within Iceland the value of NATO membership had been questioned during previous cod wars. Whereas her extensions of fishery limits had provoked conflict with some of her NATO allies, notably Britain, the eastern bloc nations had generally respected them. In the Fourth Cod War, Icelandic public opinion increasingly turned against NATO membership, particularly after the US Government turned down her request to acquire, or charter, small naval vessels to supplement her coastguard fleet—apparently partly on the grounds that they might have been used against Britain. Yet Iceland, with the US airbase at Keflavik, remained of crucial strategic importance in the unremitting Cold War and NATO would not risk its loss.

The NATO dimension proved of increasing embarrassment to Britain at a time when Iceland was gradually gaining better publicity from the world's media. Not only was world opinion swinging behind Iceland's stance but there were growing demands in Britain for her own 200-mile Exclusive Economic Zone.[44] By the middle of May 1976 Germany and Belgium had reached agreement with Iceland and both the British Government and its beleagured distant-water trawling industry were increasingly isolated.

As the tide turned increasingly in Iceland's favour, the British Government searched desperately for a way of reactivating negotiations. After preliminary approaches through the medium of NATO allies, the British Foreign Secretary, Anthony Crosland, MP for Grimsby, asked for a meeting with Icelandic Ministers and this took place in Oslo. Further meetings were held between 31 May and 2 June 1976, until an agreement was finally reached and diplomatic relations restored. Under the terms of the Oslo Agreement Britain was allowed to send twenty-four trawlers out of a list of ninety-three into Icelandic waters at any one time.

As in previous conflicts the balance of power in terms of vessels and crews deployed bore little relationship to the eventual outcome. Iceland had deployed two helicopters and one Fokker Friendship aircraft in addition to its six small coastguard vessels and two chartered trawlers, crewed in all by 170 men. In contrast, Britain used at one time or another twenty-two frigates, totalling 39,850 tons and 5,220 crew, in addition to seven supply vessels and nine ocean-going tugs complete with crews. Aerial reconnaissance was assisted by the use of modern RAF Nimrods and one estimate puts the Cod War's cost to the British taxpayer in the region of £40 million.[45] The bill for repairing frigates damaged in collisions was probably more than £1 million alone.[46] The total value of fish caught by British vessels in Icelandic waters in 1975 amounted to little more than £29 million. Before the last Cod War, Iceland had offered Britain 65,000 tons a year but afterwards Britain settled for 50,000 tons per annum. The distant-water trade called the agreement a sell-out and were highly critical of Foreign Secretary Anthony Crosland. In fact the agreement reached was scheduled to run for just six months. When it expired on 1 December 1976 it was not renewed[47] and British trawlers were obliged to leave Icelandic waters for good.

Epilogue

Despite darkening clouds, the British trawling trade still possessed a feeling of vigour in the early 1970s. In its report on 1970 the White Fish Authority noted that the greater profitability of the distant-water fleet had produced an upsurge in orders for new vessels, particularly freezers. Hull and Grimsby, the premier distant-water ports, still accounted for almost half of British fish landings and no port in Europe came close to matching their levels. In 1973 the trade still did well, as auction prices—up forty-five per cent for a ten stone kit of cod on the Humber over the year—rose even faster than the nation's febrile inflation rate.[1]

But the conflicts with Iceland and the effects of the Common Fisheries Policy were not the only afflictions sapping at the very vitals of the distant-water trawling trade. The OPEC rises in the price of fuel oil, which began to take effect from the end of 1973, not only pushed the western world into a deep recession but thrust the distant-water trawling trade, already trying to embrace rapid and unprecedented change, towards the edge of extinction. Oil for voyages to distant-water grounds was a critical factor in maintaining profitability and, as a gallon of fuel which had cost seven pence in 1973 rocketed to twenty-one pence by March 1975, costs went through the ceiling. The fuel bill for the whole UK fishing industry rose from £8 million in 1972 to over £33 million in 1976.[2]

For a short while a number of the larger companies were shielded by their existing contracts with the oil companies but this afforded little more than a breathing space. The first casualties were the 45 remaining oil-burning steam trawlers. They rapaciously consumed heavy oil and by mid-1974 each vessel faced an annual fuel bill in the region of £100,000. Many of these vessels, often over twenty years of age, had been kept in operation during the prosperous years of 1972 and 1973 but in 1974/5 virtually all were withdrawn from service. The British distant-water fleet could boast 122 side-winders and 46 freezer vessels at the beginning of 1974[3] but by the end of 1975 only 89 side-winders and 44 freezers remained in operation, a twenty-seven per cent decrease. The middle-water fleet shrank from 171 vessels to 144 during the same period.[4]

The rise in oil prices was not only reflected in soaring vessel fuel costs. Much trawling gear was by then manufactured from oil-based products; for example, synthetic fibres used for making nets and ropes. Onshore, distribution costs were also subjected to large rises. At the same time as it experienced the full brunt of runaway cost inflation in the spring of 1974, the trade faced a downturn in its market. Depression prevailed throughout the fish markets of western Europe caused largely by the very high level of stocks being held in cold storage. It was not until such supplies began to run down in the following year that any degree of buoyancy returned to the market.

After losing the Icelandic grounds in 1976, Hull, Grimsby and Fleetwood redeployed their shrinking distant-water fleets, with the former two ports increasing their exploitation of the Norwegian and North Sea grounds. But traditional options were decreasing as fishing limits extended and the middle- and distant-water fleets were excluded from the Barents Sea, Faroes, Norwegian coast and Newfoundland grounds. Aberdeen was to suffer particularly from the loss of the Faroese grounds. Soon large fleets of sophisticated fish catchers had few places to turn. New ventures were tried: blue whiting was sought off the western edge of the continental shelf and the large freezer vessels turned their hand to the south-western mackerel fisheries. In many respects this was all a case of too little too late. Such activities were unable to sustain the large fleet and the trade which depended on it.

Ships and crews were not the only sectors affected. Reduced landings from traditional grounds left processors and merchants with a sometimes desperate search for supplies. Their reduced throughput led them to reduce their workforce. The continued reduction in the size of the fleet meant progressively less and less work for the ship repairers and allied trades. The construction of new deep-sea vessels all but dried up and this had tremendous repercussions for the shipyards which traditionally built them.

As late as 1976 Hull remained the leading British port by value, but within six years its landings had fallen to 15,000 tonnes and it no longer ranked amongst the leading British fishing ports.[5] Grimsby fared a little better having retained a considerable interest in the North Sea, but its distant-water fleet was decimated. Fleetwood had abandoned its distant-water fishing by the early 1980s. Though prices for fish had recovered, and the market exhibited a greater degree of buoyancy, costs continued to rise and the decreasing opportunities to fish meant catches continued to decline. By June 1978 the British distant-water fishing fleet was shrinking at the rate of two vessels a week. The laying up and scrapping continued unabated throughout the remainder of the 1970s. Many traditional side-winders were scrapped—most of those belonging to Hull were broken up in Drapers Yard near Sammy's Point on the Humber. Some were converted for use as oil rig standby vessels. The Hull trawlers *St Giles*, owned by Thomas Hamling and Co, and *Westella*, owned by J. Marr and Son, were sold to Greenpeace, the latter being scuttled off Spain in 1979 after a collision with a whaler. Many freezers were also disposed of over the next few years. The *Arctic Buccanner* and *Arctic Galliard* moved with their crews to New Zealand to work out of the port of Dunedin as the *Otago Buccaneer* and the *Otago Galliard.* Others went to South Africa, Australia, Norway, Nigeria and Iran. Not all continued to fish. Some were converted into survey or standby vessels. Five of the remaining Hull freezers, *Junella, Cordella, Northella, Farnella* and *Pict* were requisitioned during the Falklands War for minesweeping duties.[6]

The dramatic and absolute decline of the deep-water sectors was accompanied by a shift in the gravity of the industry. Though the whole industry was drastically affected by the upheavals of the 1970s, the inshore sector proved much more adaptable, particularly in Scotland. For most of the twentieth century the British fishing industry had been dominated by the English and Welsh deep-sea sectors and this was still the case at the end of the 1960s. However, by 1980 the Scottish inshore ports, Peterhead in particular, were at the leading edge of a much altered trawling and fishing industry.

With the benefit of hindsight it may appear obvious that, from the 1950s at least, the long-term vitality of Britain's traditional distant-water fleet was at best uncertain given deepening international worries about conservation, a growing post-war interest by many coastal states in the economic value of their adjacent waters, and the growing challenge to the concept of freedom of the seas. However, it might also be argued that the dramatic pace and wide-ranging series of events which overwhelmed the distant-water fisheries in the 1970s compressed so much change into so short a period of time that few industries could have expected to adapt in such circumstances. Yet even if those changes had occurred over a longer timespan one must seriously question whether many firms in the sector had really anticipated the radical changes in strategy they were called upon to embrace. The traditional response of many distant-water trawler owners, since at least the time of Huxley, to shortages of fish supplies was to build increasingly efficient vessels and seek out new grounds even further afield. A principal duty of government in their eyes was to protect their 'rights' on distant-water grounds. Although there was a growing awareness of the need for conservation in the twentieth century, in practical terms the catching efficiency of the world's trawling fleets continued to outstrip the resource base. During the 1960s, with worries about catch levels on existing grounds, the British industry embarked upon the construction of a new class of vessel—the stern freezer: even more efficient, and able to work on grounds in international waters at or beyond the very limit of the old side-winder trawlers. The vessels may have been new, highly efficient and embraced the latest technological developments but the concepts which underlay the decision to construct them can be traced back to the nineteenth century.

Perhaps the distant-water sector might have been able to secure a stronger, albeit different, stake in the future had it been part of a more united fishing industry. In practice, however, the trade had never been easy to unite. The inshore and deep-sea sectors had a history of poor co-operation and many distant-water owners retained an arrogant management style and attitude to unions which was hardly conducive to the development of good industrial relations. During a decade when the industry was being assailed from all directions, it had little experience of presenting a united front in the pursuit of its interests.

Today, the British distant-water sector can scarcely be described as even a shadow of its former self. But the world's fishing activities have continued to expand. There is in theory a realization of the need to take account of the maximum sustainable yield which can be taken from fish stocks, and a greater awareness of other aspects of human activity which threaten this unique maritime resource, particularly the continuous pollution of the seas and the destruction of coastal habitats. But in practice, the world often continues to act in a way which appears to treat fishing as an extractive industry. The annual global fisheries catch, which was three million tonnes around 1900, peaked at over 82 million tonnes in 1989.[7] The world's heavily subsidized fishing fleet, has doubled in size since 1970. Reports show that all the world's largest fishing areas are being fished at or above their natural limits and more than half are suffering a serious decline in catches. Politicians, fishermen and scientists argue about the level of the maximum sustainable yield which can be taken from various stocks and about which conservation strategies are the most effective: ground closure, a moratorium on specific species, taking boats out of service, the use of nets with larger mesh or restricting fishing days. Apparently sound practices introduced by some states are often undone by the activities of others. Moreover, there is a widespread distrust amongst fishermen of many nations regarding the motives of their foreign counterparts and their maintenance of agreed conservation measures. In the 1990s the worldwide catching capacity greatly exceeds the sustainable marine catch. This level of effort is maintained with the assistance of governments who continue to subsidize the industry to the tune of $54 billion worldwide. With few new grounds left to open up, fishing fleets have more recently turned to lower-value species which were once considered not always worth the taking. The 1995 Canadian/Spanish dispute concerned such a species.

The sheer pace of the British distant-water fleet's decline left many of its trawlermen high and dry. Some went into North Sea trawling or coastal fisheries, other worked on oil rig standby vessels, whilst large numbers took jobs ashore—trawlermen being especially valued in continuous-process industries with shift work and long and unsociable hours. But many were left unemployed as the country was gripped by recession. A unique labour force was dispersed and the skills and experiences it collectively retained scattered to the winds. Unlike workers in many other traditional industries which were shedding labour, the trawlermen did not always receive the benefit of financial compensation for losing their livelihoods. Classed as casual workers, Hull and Grimsby trawlermen with twenty or more years of service to the industry were to fight into the 1990s for redundancy payments. Rough justice for a resilient labour force which had endured and given so much.

Appendix

Demersal fish landings in Britain by British vessels (cwts)

Year	*England & Wales*	*Scotland*	*Total*
1912	8,749,591	2,862,586	11,612,177
1913	8,360,769	2,639,642	11,096,021
1921	7,867,037	2,735,252	10,506,679
1922	9,007,004	2,828,945	11,835,949
1923	8,121,970	2,069,160	10,191,130
1924	8,569,270	2,310,981	10,880,251
1925	9,287,404	2,468,208	11,755,612
1926	8,556,946	2,404,281	10,961,227
1927	9,307,968	2,619,319	11,927,287
1928	9,016,620	2,553,212	11,569,832
1929	9,732,270	2,591,612	12,323,882
1930	11,454,125	2,692,620	14,146,745
1931	11,109,473	2,778,759	13,888,232
1932	11,143,727	2,795,622	13,939,349
1933	10,941,278	2,680,849	13,622,127
1934	10,920,390	2,479,078	13,399,468
1935	11,396,644	2,395,683	13,792,327
1936	12,637,573	2,275,711	14,913,284

1937	13,333,347	2,481,731	15,815,078
1938	12,644,000	2,976,000	15,620,000
1945	5,134,488	2,109,427	7,243,915
1946	10,768,380	3,090,121	13,858,501
1947	11,653,665	3,635,281	15,288,946
1948	11,915,000	3,357,011	15,272,011
1949	12,571,093	3,159,566	15,730,659
1950	10,900,275	2,997,633	13,897,908
1951	12,290,000	2,982,191	15,272,191
1952	12,120,370	3,090,420	15,210,790
1953	11,536,722	3,185,124	14,721,846
1954	11,242,104	3,257,355	14,499,459
1955	11,878,927	3,562,745	15,441,672
1956	11,798,889	3,750,855	15,549,744
1957	10,612,838	4,115,844	14,728,682
1958	10,633,224	4,074,794	14,708,018
1959	10,329,465	3,692,609	14,022,074
1960	10,239,000	3,367,000	13,606,000
1961	9,440,000	3,364,000	12,804,000
1962	10,239,000	3,185,000	13,424,000
1963	9,554,000	3,648,000	13,202,000
1964	9,243,000	4,208,000	13,451,000
1965	9,606,000	4,764,000	14,370,000
1966	9,428,000	4,582,000	14,010,000
1967	9,617,000	4,268,000	13,994,000
1968	9,929,000	4,327,000	14,256,000
1969	10,165,000	4,069,000	14,234,000
1970	9,381,000	4,933,000	14,314,000
1971	8,628,000	5,372,000	14,000,000

Rounded to nearest thousand from 1960.

Source: Sea Fisheries Statistical Tables.

Notes

Chapter One

1. BPP, 1803, X, 138–9, S.C. on British Fisheries, *Report*,
2. BPP, 1785, VII, 7, S.C. on British Fisheries, *Third Report*.
3. Gray, M., *The Fishing Industries of Scotland 1790–1914* (Aberdeen 1979) pp. 52–3.
4. BPP, 1798, S.C. on British Fisheries; BPP, 1803, X, 130–5, *Report*.
5. The British Fisheries Society was founded in 1786 with the support of Parliament and through the activities of a band of public spirited leaders. Its aim was to foster economic activity, particularly fishing, in the west and north of Scotland. For full details see J. Dunlop, *The British Fisheries Society 1786–1893* (Edinburgh 1981).
6. RHW, AF1/5. 1 Jan. 1825.
7. Waites, B., 'The Medieval Ports and Trade of North East Yorkshire' *Mariner's Mirror*, 63 (1977), 144–6.
8. RHE, AF4/2, 15 Nov.1833.
9. BPP, 1785, VII, 14–15, S.C. on British Fisheries, *Third Report*.
10. BPP, 1817, XIV, 383, Papers Relating to Salt Duties.
11. Young, G., *A History of Whitby and the Vicinity*, vol 2 (Whitby 1817) pp. 820–3.
12. BPP, 1866, XVII–XVIII, qq. 6560–6, R.C. on Sea Fisheries, *Minutes of Evidence*.
13. BPP, 1836, XXXII, 149, E.C. on Irish Fisheries, *Appendix*.
14. BPP, 1785, VII, 14–15, S.C. on British Fisheries, *First Report*.
15. BPP, 1785 VII, 4–5, S.C. on British Fisheries, *Report into Pilchard Fishery*.
16. Robinson, R., *A History of the Yorkshire Coast Fishing Industry 1780–1914* (Hull 1987) p. 19.
17. Gray, *Fishing Industries*, pp. 42–3.

18. Northway, A. M., 'The Devon Fishing Industry 1760–1860' (MA thesis, University of Exeter, 1970) p. 12
19. Robinson, *History*, p. 28.
20. BPP, 1817, XIV, 383, Papers Relating to Salt Duties.
21. RHE, AF1/27, 6 Nov. 1820 and 8 Feb. 1822; AF1/9, 9 Oct. 1832.
22. BPP, 1797/8, XX, 130–1, S.C. on British Fisheries, *Report*.
23. BPP, 1817, XIV, 383–5, Papers Relating to Salt Duties.
24. Robinson, R., 'The English Fishing Industry 1790–1914: A Case Study of the Yorkshire Coast' (PhD thesis, University of Hull, 1985) p. 56.
25. 41 Geo III, Cap. 99.
26. Robinson, R., 'The Fish Trade in the Pre-Railway Era: The Yorkshire Coast 1780–1814' *Northern History*, XXV (1989), 230.
27. NYCRO, *Scarborough Letters*, 23 March 1813.

Chapter Two

1. Alward, G. L., *The Sea Fisheries of Great Britain and Ireland* (Grimsby 1932) p. xx
2. *Fisheries Exhibition Literature* Vol IV (1883) p. 11.
3. 1 Geo I, 2 Cap XVIII.
4. HCRO, NEDSFC, 13 Oct. 1896.
5. Anson, P., *Fishermen and Fishing Ways* (1932, repub. Wakefield 1975) p. 122.
6. BPP, 1906, II, Report of Delegates Attending Meetings of the International Council for the Exploration of the Sea in 1903–5.
7. Northway, *thesis*, p. 57
8. Robinson, *History*, p. 10
9. Northway, *thesis*, p. 57
10. BPP, 1800, X, 130–1, S.C. on British Herring Fisheries, *First Report*.
11. Northway, *thesis*, p. 57
12. Northway, *thesis*, p. 12.
13. Chaloner, W. H., 'Trends in Fish Consumption' in Barker, T. C., Barker, J. C. and Yudkin, J., eds, *Our Changing Fare* (1966) p. 103.
14. BPP, 1833, XIV, qq. 2189–90, S.C. on British Channel Fisheries, *Minutes of Evidence*.
15. BPP, 1836, XXXII, qq. 242–6, R.C. Irish Fisheries, *Minutes of Evidence*.
16. Northway, *thesis*, p. 168.
17. Robinson, *thesis*, pp. 138–43.
18. *Yorkshire Gazette*, 30 July 1831.
19. *Hull Rockingham*, 9 June 1832.

20. *Yorkshire Gazette*, 17 Sept. 1833.
21. Robinson, *History*, p. 46.

Chapter Three

1. Reussner, G., 'The Whitby and Pickering Railway, Income and Traffic', *Moors Line* (Spring 1981) 15–17.
2. *Hansard*, 20 March 1845.
3. ibid.
4. Robinson, R., 'The Evolution of Railway Fish Traffic Policies, 1840–66' *Journal of Transport History*, 7 (1986), 33–4.
5. Bagwell, P. S., *The Railway Clearing House* (1968) pp. 21–2.
6. *Hansard*, 20 March 1845.
7. *Hull and Eastern Counties Herald*, 27 Jan. 1842 and 3 March 1842.
8. Robinson, R., 'Evolution', 36–7.
9. PRO, RAIL 1080/162, 20 June 1850.
10. PRO, RAIL 318.1, 13 Jan. 1857.
11. Robinson, *thesis*, pp. 109–113, 129.
12. Robinson, *thesis*, p. 126.
13. Gillett, E. and MacMahon, K. A., *A History of Hull* (Hull 1980) pp. 306–11.
14. B. Thomas, *Migration and Economic Growth* (2nd Ed. Cambridge 1972) p. 73.
15. Mayhew, H., *London Labour and the London Poor* (1861–2, repr. 1967) pp. 165–70.
16. Cutting, C. L., *Fish Saving* (1955) pp. 240–1.

Chapter Four

1. Northway, *thesis*, pp. 81–3.
2. Robinson, *History*, p. 47.
3. Bellamy, J. M., 'Some Aspects of the Economy of Hull in the Nineteenth Century With Special Reference to Business History' (PhD thesis, University of Hull, 1965) p. 144.
4. Bellamy, J. M., 'Pioneers of the Hull Trawl Fishing Industry' *Mariner's Mirror*, 51 (1965), 185.
5. Robinson, R., 'The Rise of Trawling on the Dogger Bank Grounds: The Diffusion of an Innovation' *Mariner's Mirror*, 75 (1989), 83–4.
6. *Leeds Mercury*, 20 Jan. 1845.
7. *The Times*, 4 Feb. 1845.
8. Robinson, *thesis*, pp. 145–9.
9. BPP, 1866, XVII–XVIII, qq. 6012–16, R.C. on Sea Fisheries, *Minutes*.

10. BPP, 1849, LI, 22, Captain Washington's Report on Fishing Vessels, *Appendix.*
11. NYCRO, Scarborough Custom House Vessel Registers, 1850–60.
12. Robinson, *thesis*, pp. 361–94.
13. Northway, *thesis* pp. 161–8.
14. Alward, G. L., *The Sea Fisheries of Great Britain and Ireland* (Grimsby 1932) pp. 336–42.
15. Robinson, *thesis*, pp. 461–5.
16. NYCRO, Whitby Custom House Vessel Registers, 1849–69.
17. BPP, 1890, VIII, Sea Fisheries (England and Wales) Inspectors' Report.
18. HRO, Hull Custom House Vessel Register, 1844, no. 4.
19. HRO, Hull Custom House Vessel Registers, 1844–9.
20. *Hull Advertiser*, 1 August 1863.
21. BPP, 1866, XVII–XVIII, qq. 7788–92, R.C. on Sea Fisheries, *Minutes of Evidence.*
22. BPP, 1878/9, XVII, 111, Sea Fisheries in England and Wales, *Minutes of Evidence.*
23. *Eastern Morning News*, 16 Jan. 1884.
24. Bellamy, J. M., *The Trade and Shipping of Nineteenth-Century Hull*, (Hull 1970, repr. 1979) p. 38.
25. Alward, *Sea Fisheries*, p. 321.
26. Malster, R., *Lowestoft East Coast Port* (Lavenham 1982) pp. 105–9.
27. *Hull and Eastern Counties Herald*, 16 June 1855.
28. Gillett, E., *A History of Grimsby* (Hull 1970, repr. 1986) pp. 230–1.
29. BPP, 1866, XVII–XVIII, qq. 16073 and 16209, R.C. on Sea Fisheries, *Minutes of Evidence.*
30. Gillett, *Grimsby* pp. 231–2.
31. Bellamy, *Trade and Shipping*, p. 37.
32. HRO, Hull Custom House Vessel Registers, 1850–1860.

Chapter Five

1. BPP, 1866, XVII–XVIII, xi–xv, R.C. on Sea Fisheries, *Report.*
2. This body administered the British herring fisheries from 1809 and the cod, ling and hake fisheries from 1809 until 1850, after which its activities were confined to Scotland. Its original title was the Board of British Herring Fisheries, and its records are deposited in the Register House, Edinburgh under the classification AF.
3. BPP, 1856, LIX, 185–6, Report to the Treasury by Mr J. G. Shaw Lefevre on the Fishery Board 1849.
4. BPP, 1866, XVII–XVIII, q. 6875, R.C. on Sea Fisheries, *Minutes of Evidence.*

5. *Whitby Gazette*, 13 Dec. 1862 and *Scarborough Gazette*, 15 Jan. 1863.
6. *Hull Advertiser*, 1 April 1863.
7. Barback, R. H., *The Political Economy of the Fisheries* (Hull 1966) pp. 18–19.
8. BPP, 1866, XVII–XVIII, q. 12415, R.C. on Sea Fisheries, *Minutes of Evidence*.
9. Robinson, *thesis*, pp. 239–40.
10. Barback, *Political Economy*, pp. 18–19.
11. *Annual Statement of Trade and Navigation for 1878*.
12. BPP, 1882, XVII, viii, Board of Trade Committee on Relations between Owners, Masters and Men, *Report*.
13. Gillett, E., *History*, p. 247.
14. Boswell, D., *Sea Fishing Apprentices of Grimsby* (Grimsby 1974) pp. 28–9.
15. BPP, 1882, XVII, q. 3028, Board of Trade Committee on Relations between Owners, Masters and Men, *Minutes of Evidence*.
16. BPP, 1894, LXXIX, 4–6, Report by A. D. Berrington and J. S. Davy of Investigation into the Fishing Apprenticeship System.
17. BPP, 1882, XVII, q. 2434, Board of Trade Committee on Relations between Owners, Masters and Men, *Minutes of Evidence*.
18. BPP, 1882, XVII, q. 199, Board of Trade Committee on Relations between Owners, Masters and Men, *Minutes of Evidence*.
19. BPP, 1883, XVIII, 7–8, Report of Board of Trade on System of Deep Sea Trawl Fishing in the North Sea.
20. Boswell, *Apprentices*, p. 65.
21. *Hull Advertiser*, 28 Dec. 1864 and 7 Jan. 1865.
22. *Hull and Eastern Counties Herald*, 11 and 18 Sept. 1873.
23. *Hull and Eastern Counties Herald*, 16 April 1864.
24. BPP, 1882, XVII, q. 225, Board of Trade Committee on Relations between Owners, Masters and Men, *Minutes of Evidence*.
25. BPP, 1882, XVII, q. 6457, Board of Trade Report on Relations between Owners, Masters and Men, *Minutes of Evidence*.
26. BPP, 1882, XVII, q. 545, Board of Trade Report on Relations between Owners, Masters and Men, *Minutes of Evidence*.
27. Quoted by Boswell, *Apprentices*, pp. 85–6.
28. BPP, 1882, XVII, q. 5331, Board of Trade Committee on Relations between Owners, Masters and Men, *Minutes of Evidence*.
29. Boswell, *Apprentices*, pp. 85–6.
30. *Grimsby News*, 11 Jan. 1878.
31. BPP, 1894, LXXIX, 4–6, Report by A. D. Berrington and J. S. Davy of Investigation into Fishery Apprenticeship System,
32. Gillett, *History*, p. 261.

33. *Eastern Morning News*, 6 Mar. 1882; 20 Mar. 1882; 24 Mar. 1882; 6 May 1882.
34. *Eastern Morning News*, 13 July 1882.
35. *Eastern Morning News*, 2 Sept. 1882.
36. BPP, 1894, LXXIX, 4–6, Report by A. D. Berrington and J. S. Davy of Investigation into Fishery Apprenticeship System.

Chapter Six

1. Robinson, *thesis*, p. 164.
2. BPP, 1878–9, XVII, 102, Sea Fisheries of England and Wales, Report of F. Buckland and S. Walpole, *Minutes of Evidence*.
3. BPP, 1866, XVII–XVIII, qq. 6873–5, R.C. on Sea Fisheries, *Minutes of Evidence*.
4. NYCRO, Scarborough Custom House Vessel Register, 22 July 1862.
5. Dade, D., 'Trawling Under Sail on the North East Coast', *Mariner's Mirror*, 18 (1932), 363–5.
6. Robinson, *thesis* p. 384.
7. *Hull and Eastern Counties Herald*, 23 Jan. 1868.
8. BPP, 1882, XVIII, xi, Board of Trade Committee on Relations Between Owners, Master and Men, *Report*.
9. *Grimsby News*, 10 Mar. 1877.
10. *Hull and Eastern Counties Herald*, 30 Sept. 1869.
11. Alward, *Fisheries*, pp. 145–6
12. *Eastern Morning News*, 12 Sept. 1868.
13. *Scarborough Gazette*, 1 Jan. 1886.
14. *Grimsby News*, 30 Oct. 1878.
15. BPP, 1836, XXXII, q. 2426, R.C. on Irish Fisheries, *Minutes of Evidence*.
16. BPP, 1883, XVIII, 3–5, Report of Board of Trade on System of Deep Sea Trawl Fishing in the North Sea.
17. *Grimsby News*, 18 May 1879.
18. *Eastern Morning News*, 12 April 1880.
19. *Eastern Morning News*, 28 June 1884.
20. *Eastern Morning News*, 17 April 1880.
21. *Grimsby News*, 4 Oct. 1878.
22. *Grimsby News*, 28 Jan. 1876.
23. *Grimsby News*, 4 Oct. 1878.
25. *Grimsby News*, 3 Sept. 1880, 10 Sept. 1880 and 1 Oct. 1880.
26. Gillett, *History*, p. 267.
27. *Hull and Eastern Counties Herald*, 16 April 1852.
28. *Hull Advertiser*, 18 Oct. 1856.
29. *Eastern Morning News*, 22 Oct. 1880.

30. *Eastern Morning News*, 6 Nov. 1880.
31. Brown, R. M., *Waterfront Organisation in Hull* (Hull 1972) pp. 24–5.
32. *Scarborough Gazette*, 6 May 1880.
33. *Eastern Morning News*, 28 June 1884.
34. BPP, 1881 LXXXII, Report of W. H. Higgins Esq., QC on the Outrages Committed by Foreign upon British Fishermen in the North Sea.
35. BPP, 1881, LXXXII, Report of W. H. Higgins Esq., QC, on the Outrages Committed by Foreign upon British Fishermen in the North Sea.
36. BPP, 1881, LXXXII, Report of W. H. Higgins Esq., QC, on the Outrages Committed by Foreign upon British Fishermen in the North Sea.
37. *Hansard*, 12 June 1883.
38. Robinson, *thesis*, pp. 253–6.
39. Villiers, A., *The Deep Sea Fishermen* (1970) pp. 89–95.
40. BPP, 1883, XVIII, 7–8, Report of the Board of Trade into the System of Deep Sea Trawl Fishing in the North Sea.
41. Gillett and MacMahon, *Hull*, pp. 317–18.
42. *Eastern Morning News*, 9 Nov. 1883.
43. *Eastern Morning News*, 8 Jan. 1886.
44. *Eastern Morning News*, 8 Jan. 1886.
45. *Scarborough Gazette*, 14 April 1887.

Chapter Seven

1. Bellamy, *thesis*, pp. 328–30.
2. *Grimsby News*, 18 Jan. 1877.
3. Tunstall, J., *The Fishermen* (1962) pp. 17–18 and G. Morey, *The North Sea* (1968) p. 130.
4. Von Tunzelman, G. N, *Steam Power and British Industrialisation to 1860* (Oxford 1978) p. 278.
5. *Hull Advertiser*, 18 Nov. 1853.
6. BPP, 1866, XVII–XVIII, qq. 314–16, 29764–70, 39175–39210, 39886–7, R.C. on Sea Fisheries, *Minutes of Evidence*.
7. I am indebted to Harry S. Taylor for additional information on early steam-trawling ventures. See also *Torquay Directory and South Devon Journal*, 13 May 1878.
8. BPP, 1879 XVII, xviii, Sea Fisheries of England and Wales, report of F. Buckland and S. Walpole, *Report*.
9. Anon., *The Origin of the Tyne Lifeboat Service and of the Tynemouth Volunteer Life Brigade* (North Shields 1928) pp. 5–11.

10. BPP, 1879, XVII, 128, Sea Fisheries of England and Wales, report of F. Buckland and S. Walpole, *Minutes of Evidence.*
11. *Shields Daily News*, 3 April 1878.
12. *Grimsby News*, 18 Jan. 1878.
13. *Whitby Gazette*, 24 Dec. 1880.
14. BPP, 1879, XVII, 105, Sea Fisheries of England and Wales, report of F. Buckland and S. Walpole, *Minutes of Evidence.*
15. Scarborough Custom House Vessel Register, 3 Aug 1878.
16. Godfrey, A., *Yorkshire Fishing Fleets* (Clapham 1974) p. 124.
17. *Scarborough Gazette*, 19 Jan. 1882.
18. Scarborough Public Library, Maritime Papers and Letters.
19. Robinson, *thesis*, p. 319–24.
20. *Scarborough Gazette*, 25 Nov. 1897.
21. Robinson, *thesis*, pp. 319–24.
22. *Grimsby News*, 16 Dec. 1881.
23. *Grimsby News*, 28 Feb. 1890.
24. Bellamy, *thesis*, pp. 330–2.
25. Gray, *Fishing Industries*, pp. 166–7.
26. Aberdeen Custom House, Fishing Vessel Register, 1875.
27. Aberdeen Custom House Vessel Register, 28 Feb. 1881.
28. Aberdeen Custom House Vessel Register, 20 Mar. 1882.
29. Aberdeen Custom House Vessel Register, 1883–1890.
30. Gray, *Fishing Industries*, p. 171.
31. *Aberdeen Journal*, 12 July 1890.
32. BPP, 1885, XVI, qq. 1869–1913, R.C. on Trawling, *Minutes of Evidence.*
33. *Scarborough Gazette*, 13 Sept. 1900.
34. Gray, *Fishing Industries*, p. 169.
35. Gray, *Fishing Industries*, p. 166.
36. *Eastern Morning News*, 17 Dec. 1885.
37. *Eastern Morning News*, 12 Jan. 1888 and 2 Feb. 1888.
38. Harley, C. K., 'The Shift from Sailing Ships to Steam Ships 1850–1890' in D. M. McCloskey, ed., *Essays on a Mature Economy: Britain after 1840*, pp. 219–20.

Chapter Eight

1. Saul, S. B., *The Myth of the Great Depression* (1969) pp. 34–5.
2. Gray, *Fishing Industries*, pp. 146–9.
3. Robinson, *thesis*, pp. 278–84.
4. BPP, 1866, XVII–XVIII, R.C. Sea Fisheries, *Report.*
5. *Fisheries Exhibition Literature* (1883) Vol. IV, p. 14.

6. Garstang, W., 'The Impoverishment of the Sea' *Journal of the Marine Biological Association*, VI (1900) 1–5.
7. *Aberdeen Journal*, 11 Dec. 1884 and 12 Dec. 1884.
8. *Aberdeen Journal*, 3 Mar. 1885.
9. BPP, 1879, XVI, qq. 10752–10760, Report on English and Welsh Sea Fisheries, *Minutes of Evidence.*
10. BPP, 1885, XVI, q. 8651, R.C. on Trawling, *Minutes of Evidence*
11. BPP, 1879, XVII, 111–13. Sea Fisheries of England and Wales, report of F. Buckland and S. Walpole, *Minutes of Evidence.*
12. BPP, 1885, XVI, q. 8651, R.C. on Trawling, *Minutes of Evidence.*
13. *Scarborough Gazette*, 7 April 1887.
14. *Aberdeen Journal*, 10 Mar. 1885.
15. *Eastern Morning News*, 1 April 1889.
16. Hull Central Library, *Report of the Conference of Hull and Grimsby Smackowners*, 30 April 1890.
17. Jenkins, J. T., *Sea Fisheries* (1920) pp. 151–5.
18. Aflalo, F. G., *The Sea Fishing Industry of England and Wales* (1904) p. 174; and Jenkins, J. T., *The Sea Fisheries* (1920) pp. 164–5.
19. Holm, P., 'Technology Transfer and Social Setting: The Experience of Danish Steam Trawlers in the North Sea and off Iceland', *Northern History Yearbook 1994*, pp. 137–9.
20. BPP, 1894, XV, q. 1399, S.C. on Sea Fisheries, *Minutes of Evidence.*
21. Holt, E. W. L., 'On The Icelandic Trawl Fishery', *Journal of the Marine Biological Association, 3* (1893), 129–30.
22. Thor, J. Th., 'The Beginnings of British Steam Trawling in Icelandic Waters' *Mariner's Mirror* 74 (1988), 268.
23. Holt, E. W. L., 'On The Icelandic Trawl Fishery' *Journal of the Marine Biological Association*, 3 (1893), 129–132.
24. Humber Steam Trawlers Mutual Insurance and Protecting Company, *Minutes*, 12 December 1900 and 31 December 1901.
25. Report on English and Welsh Sea Fisheries for 1903, p 5.
26. Thor, J. Th., *British Trawlers in Icelandic Waters* (Reykjavik 1992) pp. 24–7 and 54–85.
27. *The Times*, 15 April 1897.
28. BPP, 1908, XIII, qq. 8711–8750, Committee on Fishery Investigations, *Minutes of Evidence.*
29. Thor, *British Trawlers*, p. 236.
30. *The Times*, 12 April 1897.
31. Thor, *British Trawlers*, pp. 34–8.
32. Thor, *British Trawlers*, pp. 151–8.
33. Garstang, W., 'The Impoverishment of the Sea' *Journal of the Marine*

Biological Association, VI (1900); and Cushing, D. H., *The Arctic Cod* (1966), pp. 21–2.
34. Walton, J. K., *Fish and Chips and the British Working Class 1870–1940* (Leicester 1992) p. 43.
35. Robinson, R., 'The Development of the British North Sea Steam Trawling Fleet, 1877–1900', *Third North Sea History Conference 1993*, eds, Edwards, J., Holm, P., Nordvik, H., Palmer, S., and Williams, D. (Aberdeen forthcoming).
36. Scarborough Custom House Vessel Register, 22 July 1862.
37. Joensen, J. P., 'Maritime Communities in the Faroes in the Age of the Hand-Line Smack' in Fischer, L. R., Hamre, H., Holm, P. and Bruijn, J. R., eds, *The North Sea*, pp. 149–57.

Chapter Nine

1. BPP, 1894, LXXIX, 4, Report of Investigation into the Fishing Apprenticeship System, *Minutes of Evidence.*
2. Tunstall, *Fishermen*, pp. 29–34.
3. Thompson, P., Wailey T. and Lummis, T., *Living the Fishing* (1983) pp. 53–5.
4. Tunstall, *Fishermen*, p. 32.
5. Tunstall, *Fishermen*, pp. 33–4.
6. *Eastern Morning News*, 1 Jan. 1896.
7. *The Times*, 11 Jan. 1895.
8. *Eastern Morning News*, 1 Jan. 1895.
9. *The Times*, 31 Jan, 1895 and 11 Feb. 1895.
10. *Eastern Morning News*, 18 Feb. 1895.
11. *Hull Daily Mail*, 27 Jan. 1910.
12. *Hull Daily Mail*, 9 Feb. 1910.
13. *Hull Daily Mail*, 2 Jan. 1911.
14. Hough, R., *The Fleet That Had To Die* (1958) pp. 44–6.
15. Childers, E., *The Riddle of the Sands* (1903).
16. Hough, R., *The Fleet*, pp. 48–9.
17. *Eastern Morning News*, 23 Oct. 1904.
18. *The Times*, 24 Oct. 1904.
19. *The Times*, 28 Oct. 1904.
20. Hough, R., *The Fleet*, pp. 44–9.
21. Goddard, J., and Spalding, R., *Fish n' Ships* (Clapham 1987) p 21.
22. Oral testimony of William Wells, former trawler skipper, 1977.

Chapter Ten

1. Report on the Sea Fisheries, England and Wales, 1915–1918, pp. 1–4.
2. Report on the Sea Fisheries, England and Wales, 1915–1918, pp. 1–2.
3. *Scarborough Mercury*, 7 Aug. 1914.
4. *Scarborough Mercury*, 4 Sept. 1914.
5. Gill, A., *Lost Trawlers of Hull* (Beverley 1989) p. 203.
6. Godfrey, A., *Yorkshire Fishing Fleets* (Clapham 1974) p. 42.
7. Gill, *Lost Trawlers*, p. 80.
8. *Scarborough Mercury*, 14 July 1916.
9. *Scarborough Mercury*, 29 Sept. 1916.
10. Reminiscences of Herbert Johnson, Great War trawler skipper, tape recorded by his son Philip Johnson in 1980.
11. The author's great uncle, Edmund Stipetic, was lost with this trawler.
12. Report on the Sea Fisheries, England and Wales, 1915–1918, pp. 15–16.
13. Report on the Sea Fisheries, England and Wales, 1915–1918, p. 22.
14. Verbal conversation with William Wells, 1980.
15. Diary of G. F. Robinson whilst engaged upon minesweeping duties in the North Sea, 1915–1917 (unpublished).
16. Report of the Sea Fisheries, England and Wales, 1915–1918, pp. 15–16.
17. Report on the Sea Fisheries, England and Wales, 1915–1918, pp. 15–16.
18. Malster, R., *Lowestoft East Coast Port* (1982) pp. 111–12.
19. Report on the Sea Fisheries, England and Wales, 1915–1918, pp. 15–16.
20. NEDSFC, *Minutes*, 28 July 1915, 11 May 1916.
21. NEDSFC, *Minutes*, 31 Dec. 1918.
22. NEDSFC, *Minutes*, 30 June 1918.
23. NEDSFC, *Minutes*, 30 Sept. 1917.
24. Report on the Sea Fisheries, England and Wales, 1915–1918, pp. vii–viii.

Chapter Eleven

1. Report on the Sea Fisheries, 1919–1923, p. v.
2. ibid., p. vi.
3. Graham, M., *The Fish Gate* (1943) p. 152.
4. Scottish Fishery Board Reports for 1921 and 1936.

5. Report on English and Welsh Sea Fisheries, 1924–1926, p. 76.
6. *Shields Daily News*, 5 Mar. 1930.
7. Report on English and Welsh Sea Fisheries, 1919–1923, p. 9.
8. Malster, R., *Lowestoft East Coast Port* (2nd Ed., Lavenham 1987) pp. 111–12.
9. Report on English and Welsh Sea Fisheries, 1937, pp. 16–17.
10. Scottish Fishery Board Report 1934, p. 30.
11. Kelsall, R. K., Hamilton, H., Wells, F. A., and Edwards, K. C., 'The White Fish Industry' in Fogarty M. P., ed., *Further Studies in Industrial Organisation* (1948) p. 114.
12. Clarke, *thesis* p. 61.
13. *Fishing News*, 10 Nov. 1934 and 1 Dec. 1934.
14. Clarke, *thesis*, p. 55 and conversation with Christopher Gilchrist.
15. *Fishing News*, 29 Dec. 1934.
16. Sea Fish Commission for the United Kingdom, *Second Report, The White Fish Industry 1936*, p. 59.
17. Clarke, G. S., *thesis*, p. 55.
18. Cushing, D. H., *The Arctic Cod* (1966) p. 22.
19. Addy, A., 'Fifty Years of Progress in the Fishing Industry at Hull' in *Hull Association of Engineers Journal*, Vol 17, 1949.
20. Report on English and Welsh Sea Fisheries 1927, p. 15.
21. Thompson, P., Wailey, T., and Lummis, T., *Living The Fishing* (1983), pp. 114–15.
22. Kelsall, R. K., Hamilton, H., Wells F. A., and Edwards, K. C., 'The White Fish Industry' in Fogary M.P., ed., *Further Studies in Industrial Organisation* (1948) p. 114.
23. ibid., p. 159
24. ibid., p. 136
25. ibid., p. 153
26. *Fishing News*, 7 Sept. 1932; Sea Fish Commission for the United Kingdom, *Second Report, The White Fish Industry 1936*, p. 56.
27. Graham, M., *The Fish Gate* (1943) p. 95.
28. Report of the Imperial Economic Committee on the Marketing and Preparation of Food, *Fifth Report—Fish*, 1927, p. 39.
29. Sea Fish Commission for the United Kingdom, *Second Report, The White Fish Industry 1936*, pp. 56–9.
30. Cutting, D. H., *The Arctic Cod* (1966) pp. 18, 27.
31. *The Times*, 17 Nov. 1932.
32. Ford, E., *Nation's Sea Fish Supply* (1937) pp. 11–12; and Graham, M., *The Fish Gate* (1943) pp. 103–12.

Chapter Twelve

1. Thompson, P., Wailey, T., and Lummis, T., *Living the Fishing* (1983) pp. 55–6.
2. *Eastern Morning News*, 1 Mar. 1918, 26, 27 and 28 June 1919.
3. *Eastern Morning News*, 28 June 1919.
4. *Eastern Morning News*, 20 Aug. 1919.
5. *Eastern Morning News*, 28 June 1919.
6. *Eastern Morning News*, 2 Aug. 1919.
7. *Eastern Morning News*, 20 Aug. 1919.
8. Thompson, Wailey and Lummis, *Living the Fishing*, p. 74.
9. Oral testimony of Frank Bates, trawling trade unionist 1920s–1950s.
10. Thompson, Wailey, and Lummis, *Living the Fishing*, p. 56.
11. Thompson, Wailey and Lummis, *Living the Fishing*, p. 100.
12. Oral testimony of William Wells, trawlerman and trade unionist 1910–1950.
13. The author's own grandfather was blackballed in the 1930s.
14. BPP, 1934/5, IX, 6, Report by Court of Inquiry into the Hull Fishing Industry Dispute.
15. BPP, 1934/5, IX, 6, Report by Court of Inquiry into the Hull Fishing Industry Dispute.
16. J. Tunstall, *The Fishermen* (1962) p. 38.
17. *Hull Daily Mail*, 27 Jan. 1925.
18. *Hull Daily Mail*, 12 Feb. 1925.

Chapter Thirteen

1. Hutson, H. C., *Grimsby's Fighting Fleet: Trawlers and U-boats During the Second World War* (Beverley 1990) p. 9.
2. Report on the Sea Fisheries of England and Wales, 1939–1944, pp. 8–9.
3. Report on the Sea Fisheries of England and Wales, 1939–1944, p. 31.
4. Report on the Fisheries of Scotland 1939–1948, pp. 16–17.
5. Report on the Sea Fisheries of England and Wales, 1939–1944, pp. 30–1.
6. *The Times*, 2 Oct. 1939 and 18 Mar. 1940.
7. Report on the Sea Fisheries of England and Wales, 1939–1944, pp. 10–11.
8. Leo Walmsley, *Fishermen at War* (1941) pp. 45–6.
9. ibid. pp. 47–8.
10. Report on the Sea Fisheries of England and Wales, 1939–1944, pp. 32–3.

11. Report on the Sea Fisheries of England and Wales, 1939–1944, pp. 39–40.
12. Report on the Sea Fisheries of England and Wales, 1939–1944, pp. 39–40.
13. Report on the Sea Fisheries of Scotland, 1939–1948, p. 13.
14. Report on the Sea Fisheries of England and Wales, 1939–1944, pp. 42–4.
15. Report on the Sea Fisheries of Scotland, 1939–1948.
16. Report on the Sea Fisheries of England and Wales, 1939–1944, pp. 6–7.
17. Report on the Sea Fisheries of England and Wales, 1939–1944, pp. 6–7.
18. Report on the Fisheries of Scotland, 1939–1948, p. 64.
19. Report on the Sea Fisheries of England and Wales, 1939–1944, pp. 32–3.
20. P. Lund and H. Ludlam, *Trawlers Go To War* (1971) p. 1.
21. Hutson, *Grimsby's Fighting Fleet*, p. 9.
22. Lund and Ludlam, *Trawlers*, p. 2.
23. Lund and Ludlam, *Trawlers*, pp. 48–51.
24. Oral testimony of George Robinson, former trawlerman and asdic operator on *King Sol*.
25. Walmsley, *Fishermen*, pp. 216–17.
26. Lund and Ludlam, *Trawlers*, pp. 62–4.
27. Lund and Ludlam, *Trawlers*, pp. 152.
28. Lund and Ludlam, *Trawlers*, p. 250.
29. Report of the Sea Fisheries of England and Wales, 1939–1944, p. 3; and Report on the Fisheries of Scotland 1939–1948, p. 12.
30. Gill, A., *Lost Trawlers of Hull* (1989), pp. 100 and 111.
31. *The Times*, 14 Aug. 1946.

Chapter Fourteen

1. Oral testimony of William Robinson, former Hull trawlerman.
2. Tunstall, *Fishermen*, p. 48; Clarke, *thesis*, p. 74.
3. White Fish Authority, *Annual Report*, 31 March 1952.
4. White Fish Authority, *Annual Report*, 31 March 1961.
5. British Trawler Federation, *Annual Report*, p. 28.
6. White Fish Authority, *Annual Report*, 31 March 1954.
7. Kelsall, R. K., Hamilton, H., Wells, F. A., and Edwards, K. C., 'The White Fish Industry; in M. P. Fogarty, ed., *Further Studies in Industrial Organisation*, pp. 135 and 159.
8. TGWU, *Statement of Evidence Submitted to the Court of Inquiry on the Recent Trawlermen's Dispute*, 2 July 1946.

9. Tunstall, *Fishermen*, p. 176.
10. TGWU, *Statement of Evidence*, 1946.
11. TGWU, *Statement of Evidence*, Maximum Earnings, Document 13, 1946.
12. Tunstall, *Fishermen*, p. 177.
13. Final Report of Committee of Inquiry into Trawler Safety, July 1969, p. 67.
14. TGWU, *Statement of Evidence*, 1946, Document 22.
15. Tunstall, *Fishermen*, p. 69.
16. Oral testimony of Frank Bates and William Robinson.
17. TGWU, *Frank Bates Papers* 1947. Also *Hull Daily Mail*, 16 Oct. 1947.
18. Tunstall, *Fishermen*, p. 69.
19. Final Report of Committee into Trawler Safety, July 1969, p. 42.
20. Clarke, *thesis*, p. 121–2.
21. Goddard, J. and Spalding, R., *Fish n' Ships* (Dalesman 1987) p. 72.
22. Tunstall, *Fishermen*, p. 105–7.
23. The author's family, for example.
24. Final Report of Committee into Trawler Safety, July 1969, p. 79.
25. Goddard and Spalding, *Fish*, p. 79.
26. Final Report of Committee into Trawler Safety, July 1969, p. 100.
27. TGWU, *Frank Bates Papers*, 1946.
28. Conversation with Jack Wilson, former trawler skipper.
29. Final Report of Committee into Trawler Safety, July 1969, p. 100.
30. TGWU, *Additional Evidence, Subsidies and Structure*, Trawler Fishing Inquiry, 1968, p. 6.

Chapter Fifteen

1. Nicholson, J., *Food from the Sea* (1979) pp. 137–40.
2. Clarke, *thesis*, pp. 78–81.
3. *The Times*, 16 Jan. 1959. Also conversation with Bernard Stipetic.
4. Thompson, M., *Hull's Side Fishing Fleet 1946–86* (Beverley 1987) p. 38.
5. Report of the Committee of Inquiry into the Fishing Industry, 1961 Cmd 1266, p. 13.
6. Nicholson, *Food from the Sea*, p. 103.
7. Report of the White Fish Authority, 31 Mar. 1969; and *Fishing in Distant Waters*, British Trawler Federation Annual Report 1965.
8. *Fishing in Distant Waters*, British Trawler Federation Annual Report 1965.
9. ibid.
10. Warner, W., *Distant Water* (USA 1977, 1983, ed.) pp. 34–5.

11. Waterman, J. J., *Freezing at Sea: A History* (1987) pp. 48–51.
12. Warner, *Distant Water*, pp. 33–8.
13. Nicholson, *Food from the Sea*, p. 12.
14. Thompson, M., *Hull and Grimsby Stern Trawling Fleet 1961–1988* (Beverley 1988) p. 7.
15. Warner, *Distant Water*, p. 48.
16. Nicholson, *Food from the Sea*, p. 126.
17. Nicholson, *Food from the Sea*, pp. 127–8.
18. Report of the Committee of Inquiry into the Fishing Industry, 1961 Cmd. 1266, pp. 39–40.
19. Waterman, J. J., *Freezing at Sea: A History* (1987) pp. 53–69.
20. Thompson, M., *Hull and Grimsby Stern Trawling Fleet 1961–1988* (Beverley 1988) pp. 8–9.
21. Trawler Safety: Final Report of the Committee of Inquiry into Trawler Safety, July 1969, pp. 8–9.
22. See Thompson, *Hull and Grimsby Stern Trawling Fleet*.

Chapter Sixteen

1. Christy, F. T., and Scott, A., *The Commonwealth of Ocean Fisheries* (Baltimore 1965) pp. 156–60.
2. *The Times*, 25 Sept. 1951.
3. *The Times*, 19 Dec. 1951 and 8 Jan. 1952.
4. *The Times*, 8 Jan. 1952.
5. Gilchrist, A., *Cod Wars and How to Lose Them* (1978) pp. 41–2.
6. Jonsson, H., *Friends in Conflict* (1982) p. 7.
7. Jonsson, *Friends*, pp. 51–2.
8. Gilchrist, *Cod Wars*, pp. 53–4.
9. British Trawler Federation, *Annual Report*, p. 28.
10. Jonsson, *Friends*, pp. 59–60.
11. *Fish Trades Gazette*, 22 Nov. 1958.
12. Nicholson, J., *Food From the Sea* (1979), p. 171; and Jonsson, *Friends*, p. 57.
13. *The Times*, 20 Nov. 1952 and 24 Nov. 1952.
14. Jonsson, J., *Friends*, pp. 62–3; and Gilchrist, *Cod Wars*, p. 67.
15. *The Times*, 28 Nov. 1956.
16. *Fish Trades Gazette*, 18 March 1961; and Gilchrist, *Cod Wars*, p. 67.
17. Jonsson, *Friends*, p. 74.
18. *Fish Trades Gazette*, 12 April 1958; Gilchrist, *Cod Wars*, p. 67.
19. *Fish Trades Gazette*, 23 Aug. 1958.
20. *Fish Trades Gazette*, 6 Sept. 1958.
21. Gilchrist, *Cod Wars*, pp. 92–3.
22. *Fish Trades Gazette*, 6 Sept. 1958 and 4 Oct. 1958.

23. *Fish Trades Gazette*, 9 July 1960 and 30 July 1960.
24. Gilchrist, *Cod Wars*, pp. 84–8.
25. Jonsson, *Friends*, p. 95.
26. *Fish Trades Gazette*, 11 March 1961.
27. Jonnson, *Friends*, pp. 99–100.
28. Report of the Committee of Inquiry into the Fishing Industry, 1961, Cmd. 1266, p. 58.
29. *Fish Trades Gazette*, 11 Feb. 1961 and Jonnson, *Friends*, pp. 100–5.
30. Shackleton, M., *The Politics of Fishing* (1968) p. 38; and Report of the White Fish Authority, y/e 31 March 1967.
31. British Trawler Federation, *Annual Reports*, 1964 and 1965.
32. Shackleton, *The Politics*, pp. 97–9.
33. Shackleton, *The Politics*, pp. 98–100.
34. Fisheries Dispute Between UK and Iceland, 14 July 1971 to 19 May 1973, Cmd 5341.
35. Jonsson, H., *Friends*, p. 135.
36. *Fishing News International*, Oct. 1972, Vol II, No. 10.
37. *Fishing News International*, Dec. 1972 Vol II, No. 12.
38. Fisheries Dispute Between UK and Iceland, 14 July 1971 to 19 May 1973, Cmd 5341.
39. Jonsson, *Friends*, p. 140.
40. White Fish Authority, *Report*, y/e 31 March 1974, p. 56.
41. Nicholson, *Food*, pp. 172–3.
42. *Fishing News International*, Jan. 1976.
43. Jonsson, *Friends*, pp. 173–4.
44. Shackleton, *Politics*, 104.
45. Jonsson, *Friends*, p. 164 and *Fishing News International*, July 1976.
46. *Fishing News International*, July 1976.
47. White Fish Authority, *Report*, y/e 31 March 1977.

Epilogue

1. Report of White Fish Authority, year ending 31 March 1974.
2. *Fishing News International*, Vol 15, No 3, March 1976, p 21.
3. Report of White Fish Authority, year ending 31 March 1974.
4. *Fishing News International*, Vol 15, No 3, March 1976, p 20.
5. M. Shackleton, *The Politics of Fishing in Britain and France* (1986) p. 70.
6. M. Thompson, *Hull's Side Fishing Trawler Fleet 1946–1986* (1987), and *Hull and Grimsby Stern Trawling Fleet 1961–1988* (1988).
7. *The Times*, 3 August 1994.

Bibliography

PRIMARY SOURCES

Manuscripts

Aberdeen Custom House Vessel Registers: Aberdeen Custom House, Aberdeen.

British Transport Commission Records: Public Record Office, Kew.

Diary of George Francis Robinson Whilst Engaged Upon Minesweeping in the North Sea. Unpublished—copy in possession of author.

Fishery Board Records. This body was responsible for administering the British herring fisheries after 1809 and the cod, ling and hake fisheries from 1820. It oversaw a number of curing operations in England and Wales but was primarily concerned with Scotland. After 1850 the Government restricted its activities virtually to Scotland. The body was successively known as: The Commission for the Herring Fishery (1809–14) (Scotland) (1842–81), The Fishery Board for Scotland (post 1881). Its records are housed in the Register House, Edinburgh.

Frank Bates Papers. Documents accrued by Frank Bates, fishing trade unionist and revolutionary. In possession of author.

Hull Custom House Vessel Registers: Kingston upon Hull Record Office.

Humber Steam Trawlers Mutual Insurance and Protecting Company Record: Town Docks Museum, Hull.

North Eastern District Sea Fisheries Committee Records: Humberside County Record Office, Beverley, Yorkshire. A number of district sea fishery committee records dating from after circa 1890 are available in county record offices.

Scarborough Custom House Vessel Registers: North Yorkshire County Record Office, Northallerton.

Scarborough Harbour Commissioners' Minutes: Scarborough Public Library.
Whitby Custom House Vessel Registers: North Yorkshire County Record Office, Northallerton.

Parliamentary Reports and Papers (in chronological order)

S.C. on British Herring Fisheries, 1785 XVII.
S.C. on British Fisheries, 1798, 1803 X.
S.C. on British Herring Fisheries 1800 1803 X.
Papers Relating to Salt Duties, A. and P. 1817 XIV.
R.C. on Irish Fisheries, 1836 XXXII.
Captain Washington's Report on Fishing Vessel; A. and P., 1849 LI.
R.C. on Sea Fisheries, 1866 XVII–XVIII.
Report on Crab and Lobster Fisheries of England, Wales, Scotland and Ireland, 1877 XXIV.
Sea Fisheries of England and Wales, Report of F. Buckland and S. Walpole, 1878–9 XVII.
Report of W. H. Higgins Esq., QC, on the Outrages Committed by Foreign upon British Fishermen in the North Sea, 1881 LXXXII.
Board of Trade Report on Relations between Owners, Masters and Men, 1882 XVII.
Report of Board of Trade on System of Deep Sea Trawl Fishing in the North Sea, 1883 XVIII.
Report by A. D. Berrington and J. S. Davy of Investigation into the Fishing Apprenticeship System, 1893.
R.C. on Trawling 1885 XVI.
S.C. on Sea Fisheries 1894 XVI.
S.C. on Sea Fisheries, 1894, XV.
Committee on Fishery Investigations, 1908 XIII.
Report by Court of Inquiry into the Hull Fishing Industry Dispute, 1934/5 IX.
Report of Inquiry into the Fishing Industry, 1961, Cmd 1266.
Fisheries Dispute Between Britain and Iceland, 14 July 1971 to 19 May 1973, Cmd 5341.

Annual Returns of Trade and Navigation, 1868 onwards.
Annual Sea Fisheries Statistical Tables, 1886 onwards.
Reports of the Fishery Boards for Scotland.

Reports of the White Fish Authority 1951–1972

Hansard
House of Commons Journals

Theses

Bellamy, J. M. 'Some Aspects of the Economy of Hull in the Nineteenth Century with Special Reference to Business History' (Hull Univ. PhD, 1965).
Clarke, G. S. 'The Location and Development of the Hull Fishing Industry' (Hull Univ. MSc, 1957).
Northway, A. M. 'The Devon Fishing Industry 1760–1860' (Exeter Univ. MPhil, 1970).
Robinson, R. N. W. 'The English Fishing Industry 1790–1914: a case study of the Yorkshire Coast' (Hull Univ. PhD, 1985).

Newspapers/Journals

Aberdeen Journal
Eastern Morning News
Fishing News
Fishing News International
Fish Trades Gazette
Grimsby News
Hull Advertiser
Hull and Eastern Counties Herald
Hull Rockingham
Leeds Intelligencer
Leeds Mercury
Scarborough Gazette
Scarborough Mercury
Scarborough Post
Shields Daily News
The Times
Torquay Directory and South Devon Journal
Whitby Repository
Yorkshire Gazette

SECONDARY SOURCES

Place of publication is London unless otherwise stated.

Addy, A. 'Fifty Years of Progress in the Fishing Industry at Hull', *Hull Association of Engineers Journal*, vol. 17, 1949.

Alward, G. L. *The Sea Fisheries of Great Britain and Ireland*, Grimsby, 1932).

Anon. British Trawler Federation Reports.

Anon. *The Origins of the Tyne Lifeboat Service and the Tynemouth Volunteer Life Brigade* (North Shields, 1928).

Ainsworth, R. *Scarborough Guide* (Scarborough, 1844).

Anson, P. *Fishermen and Fishing Ways* (1932 repub. Wakefield, 1975).

Bagwell, P. S. *The Railway Clearing House* (1968)

Barback, R. H. *The Political Economy of the Fisheries* (Hull 1966).

Bellamy, J. M. 'Pioneers of the Hull Trawl Fishing Industry', *Mariner's Mirror*, vol. 51 (May 1965).

Bellamy, J. M. *The Trade and Shipping of Nineteenth Century Hull* (East Yorkshire Local History Series, no. 27: Hull 1971, rept. 1979).

Boswell, D. *Sea Fishing Apprentices of Grimsby* (Grimsby 1974).

Brown, R. M. *Waterfront Organisation in Hull* (Hull, 1972).

Buckley, J. *The Outport of Scarborough 1602–1843* (n.d.).

Butcher, D. *The Driftermen* (Reading 1979).

Chaloner, W. H. 'Trends in Fish Consumption', *Our Changing Fare*, eds. T. C. Barker, J. C. Mackenzie and J. Yudkin (1966).

Childers, E. *The Riddle of the Sands* (1903).

Christy, F. T. and Scott, A. *The Commonwealth of Ocean Fisheries* (Baltimore 1965).

Cushing, D. H. *The Arctic Cod* (1966).

Cutting, C. L. *Fish Saving* (1955).

Dade, E. 'Trawling under Sail on the North East Coast', *Mariner's Mirror*, vol. 18 (1932).

Dade, E. 'The Old Yorkshire Yawls', *Mariner's Mirror*, vol. 19 (1933).

Dodd, J. *The Food of London* (1856).

Dunlop, J. *The British Fisheries Society 1786–1893* (Edinburgh, 1981).

Ford, E. *Nation's Fish Supply* (1943).

Garstang, W. 'The Impoverishment of the Sea', *Journal of the Marine Biological Association* vol VI, July 1900.

Gilchrist, A. *Cod Wars and How To Lose Them* (1978).

Gill, A. *Lost Trawlers of Hull* (Beverley 1989).

Gill, A. and Sargeant, G. *Village Within a City: The Hessle Road Fishing Community of Hull* (Hull 1985).

Gillett, E. *A History of Grimsby* (Hull, 1970).

Gillett, E. and MacMahon, K. A. *A History of Hull* (Hull, 1980).
Goddard, J. and Spalding, R. *Fish n' Ships* (Clapham 1987).
Godfrey, A. *Yorkshire Fishing Fleets* (Clapham, 1974).
Goodlad, C. A. *Shetland Fishing Saga* (Shetland, 1972).
Graham, M. *The Fish Gate* (1943).
Graham, M. *Sea Fisheries: Their investigation in the United Kingdom* (1956).
Gray, M. *The Fishing Industries of Scotland 1790–1914* (Aberdeen, 1979).
Heath, P. 'North Sea Fishing in the Fifteenth Century', *Northern History*, vol. 3 (1968).
Harley, C. K. 'The Shift from Sailing Ships to Steam Ships 1850–1890' in *Essays on a Mature Economy: Britain after 1840*, ed. D. M. McCloskey.
Holm, P. 'The Modernisation of Fishing: The Scandinavian and British Model' in *The North Sea* eds. Fischer, L., Hamre, H., Holm, P. and Bruijn, J. R. (Stavanger 1992).
Holm, P. 'Technology Transfer and Social Setting: The Experience of Danish Steam Trawlers in the North Sea and off Iceland, 1879–1903', *Northern Seas Yearbook 1994*.
Holt, E. W. L. 'On the Icelandic Trawl Fishery', *Journal of the Marine Biological Association*, vol III, 1893.
Hoole, K. *A Regional History of the Railways of Great Britain Vol. IV The North East* (1965).
Horsley, P. and Hirst, A. *Fleetwood's Fishing Industry* (Beverley 1981).
Hough, R. *The Fleet That Had To Die* (1958).
Hutson, H. C. *Grimsby's Fighting Fleet* (Beverley, 1990).
Jackson, G. *Hull in the Eighteenth Century* (Hull 1972).
Jenkins, J. T. *The Sea Fisheries* (1920).
Joensen, J. P. 'Maritime Communities in the Faroes in the Age of the Hand-Line Smack', *The North Sea* eds. Fischer, L., Hamre, H., Holm, P. and Bruijn, J. R. (Stavanger 1992).
Jonsson, H. *Friends in Conflict* (1972).
Kelsall, R. K., Hamilton, H., Wells F. A., and Edwards, K. C. 'The White Fish Industry' in *Further Studies in Industrial Organisation*, ed. M. P. Fogarty (1948).
Kendall, C. *God's Hand in the Storm* (1870).
Malster, R. *Lowestoft East Coast Port* (Lavenham 1982).
March, E. *Sailing Drifters* (1952).
March, E. *Sailing Trawlers* (1953).
March, E. *Inshore Craft of Great Britain* (1970).
Mayhew, H. *London Labour and the London Poor* (1861–2, repr. 1967).
Morey, G. *The North Sea* (1968).
Michell, A. R. 'The European Fisheries in Early Modern History', *The*

Cambridge Economic History of Europe vol. 4, eds Rich, E. E. and Wilson, C. H. (Cambridge 1977).

Nicholson, J. *Food From The Sea* (1979).

Oddy, D. J. 'The Changing Techniques and Structure of the Fishing Industry' in *Fish In Britain*, eds. Barker, T. C. and Yudkin, J. (1966).

Perren, R. *The Meat Trade in Britain 1840–1914* (1978).

Reussner, G. 'The Whitby and Pickering Railway: Income and Traffic' in *Moors Line* (Spring 1981).

Robinson, R. 'The Evolution of Railway Fish Traffic Policies, 1840–66' in *Journal of Transport History*, vol 7 (March 1986).

Robinson, R. *A History of the Yorkshire Coast Fishing Industry 1790–1914* (Hull 1987).

Robinson, R. 'The Fish Trade in the Pre-Railway Era: The Yorkshire Coast, 1780–1840' in *Northern History*, vol XXV, 1989.

Robinson, R. 'The Diffusion of an Innovation: the spread of Trawling on the Northern North Sea Grounds' in *Mariner's Mirror*, vol 75 (February 1989).

Robinson, R. 'The Development of the British North Sea Steam Trawling Fleet, 1877–1900' *Third North Sea History Conference* Aberdeen 1.

Saul, S. B. *The Myth of the Great Depression* (1969).

Shackleton, M. *The Politics of Fishing in Britain and France* (1968).

Sharp, Sir C. *History of Hartlepool* (1916).

Smith, A. *An Inquiry into the Nature and Causes of the Wealth of Nations* (1776, Routledge ed. 1946).

Stern, W. N. 'The Fish Supply to Billingsgate from the 19th Century to World War II' in *Fish in Britain*, eds. Barker, T. C. and Yudkin, J.

Thomas, B. *Migration and Economic Growth* (2nd ed., Cambridge, 1972).

Thompson, M. *Hull's Side Fishing Fleet 1946–86* (Beverley 1987).

Thompson, M. *Hull & Grimsby Stern Trawling Fleet 1961–88* (Beverley 1988).

Thompson, P., Wailey, T. and Lummis, T. *Living the Fishing* (1983).

Thor, J. Th. 'The Beginnings of British Steam Trawling in Icelandic Waters', *Mariner's Mirror*, vol. 74, 1988.

Thor, J. Th. *British Trawlers in Icelandic Waters* (Reykjavik 1992).

Tunstall, J. *The Fishermen* (1962).

Von Tunzelman, G. N. *Steam Power and British Industrialisation to 1860* (Oxford 1978).

Villiers, A. *The Deep Sea Fishermen* (1970).

Waites, B. 'The Medieval Ports and Trade of North East Yorkshire' in *Mariner's Mirror*, vol. 63 (1977).

Walker, D. S. *Whitby Fishing* (Whitby, 1968).
Walton, J. K. *Fish and Chips and the British Working Class 1870–1940* (Leicester 1992).
Young, G. *A History of Whitby and the Vicinity*, vol. 2 (Whitby 1817).
Young, G. *A Picture of Whitby* (1821, 2nd ed. 1839).

Index